BOOK 1

KILLER SAMURAI SUDOKU

100 MIND-BENDING PUZZLES

100

ROBERT MACMILLAN

Introduction

Killer Sudoku is my favourite puzzle. I must have spent thousands of hours being happily frustrated by the empty squares, the rows and columns, the nonets (the 3x3 squares) and the cages with their totals. I came to love the process; first, skating around the square identifying the easy wins – looking across the rows and down the columns for that one square where a digit had to be, finding the simple cage totals – 3,4,16,17 in two cell cages, 6,7, 23, 24 in three cell cages, and so on. Then the difficult period: looking for the keys to the puzzle, seeing how a total in this row interacted with cells in that column over there, reducing the possibilities in a cell until there is only one left. And finally reaching the tipping point, when I've found enough of the key pieces, racing around the grid filling in the boxes, my satisfaction at having beaten it tempered by the slight disappointment that it was over.

Killer Samurai Sudoku magnifies all of that. It's five puzzles in one, with each of the four corner puzzles sharing a nonet with the 9x9 puzzle in the middle. Probably no individual 9x9 Killer Sudoku contains enough information to be solved independently. You must work on all the puzzles at once, connecting them together to find the solution. This is the challenge of the Samurai – being nearly five times bigger than a regular Sudoku, finding the keys is much harder. Each of the five constituent Sudokus will have its own keys, but finding them may require partially completing the other Sudokus.

The Samurai is a monster. Each puzzle is a challenge, and getting to the end is a worthy achievement. You'll know you've been in a battle and your satisfaction at the victory is earned. Enjoy it!

The rules of Samurai Sudoku

There are five regular 9x9 puzzles in the Samurai. Each 9x9 puzzle has to be solved in the usual way so that each digit 1-9 appears once and once only in each row, column and 3x3 nonet. Also, no digit can be repeated inside a killer cage. Where the 9x9 squares intersect, the 3x3 nonet at the intersection must be consistent parts of both of the 9x9 squares it is part of.

Consider this section from a regular Samurai Sudoku (ie no killer cages). Nonets A, B and C are in the top-left 9x9 Sudoku. Nonets E, F and G are in the top-right Sudoku. Nonets C, D and E form the top three rows of the centre Sudoku.

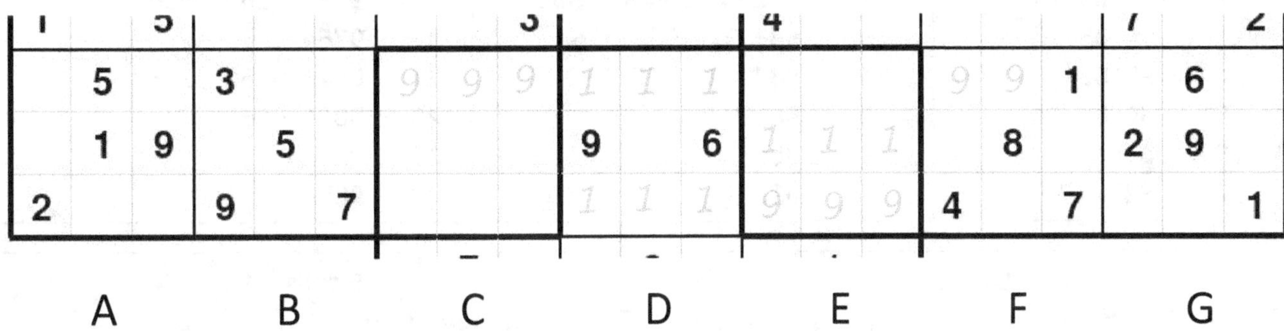

Look at the 9 digits. Nonet A contains a 9 in the middle row. Nonet B contains a 9 in the bottom row. The 9 in nonet C must therefore be in the top row. Nonets C, D and E are in the centre Sudoku. As C now has a 9 in the top row, and D already has a 9 in the middle row, the 9 in nonet E must be in the bottom row. Nonets E, F and G are in the right hand Sudoku. We've worked out that E has a 9 in the bottom row, and G already has a 9 in the middle row, so there must be a 9 in the top row of box F. This is how you can use the intersections to find more clues. Can you see how we've worked out where the 1 could be in D and E?

Before you start

This book contains 100 Killer Samurai Sudoku puzzles, providing many hours of fun, frustration and head-scratching. They are rated as 1 star to five stars similarly to the puzzles in my other Killer Suduko books. 1 and 2 star puzzles are good for those with some regular Sudoku experience who are starting out with Killer Sudoku. Experienced puzzlers may find 2 star puzzles quite straightforward and be happier with the challenges of 3 and 4 star puzzlers. Experts will find 3 star puzzles rather easy and be more challenged by 4 and 5 star puzzles. As an expert of almost two decades experience, if I complete a 5 star puzzle I feel rather pleased with myself! But be under no illusions – all of these puzzles are tricky and completing any of them is an achievement.

Finally, a request. If you enjoy this book, please help me promote it by giving it a rating on Amazon. The more ratings, the more sales, the more books I can create. Thank you!

Robert Macmillan
March 2021

Cage Combinations

2 cells Total	Combinations	3 cells Total	Combinations	4 cells Total	Combinations
3	12	6	123	10	1234
4	13	7	124	11	1235
5	14 23	8	125 134	12	1236 1245
6	15 24	9	126 135 234	13	1237 1246 1345
7	16 25 34	10	127 136 145 235	14	1238 1247 1256 1346 2345
8	17 26 35	11	128 137 146 236 245		
9	18 27 36 45	12	129 138 147 156 237 246 345	15	1239 1248 1257 1347 1356
10	19 28 37 46				
		18	981 972 963 954 873 864 765	25	9871 9862 9853 9763 9754
11	29 38 47 56				
12	93 84 75	19	982 973 964 874 865	26	9872 9863 9854 9764 8765
13	94 85 76	20	983 974 965 875		
14	95 86	21	984 975 876	27	9873 9864 9765
15	96 87	22	985 976	28	9874 9865
16	97	23	986	29	9875
17	98	24	987	30	9876

5 cells Total	Combinations	5 cells Total	Combinations
15	12345	35	98765
16	12346	34	98764
17	12347 12356	33	98763 98754
18	12348 12357 12456	32	98762 98753 98654
19	12349 12358 12367 12457 13456	31	98761 98752 98743 98653 97654
20	12359 12368 12458 12467 13457 23456	30	98751 98742 98652 98643 97653 87654

1

2

3

11			12	9		23						10			14	18			11	
11	10	13		9			6					9				23			4	11
			3	13	22			11				15		12		14	7			
17		14			10		15					24					8		9	
11			16			12						14					13	20		
		12			9		9					10			16			16		
26	17					12		13	8		15	13		14		9	8			14
	12		12	12		7	14						15		9			8		
				3				11		10		18	9			21			5	
						26				3										
						10		13		8		13		8						
						4	7		24		14									
9	15		15	6			16	6			15			9		12	9	10		10
	10				9			5		10	6		17		4			13		
15	9		12		13		11				24					13	17	11		13
	15		8	20			9					7		11						
	18				18			7				15		11	7		14			13
			24		12							11			14			13	10	
	19	9		15		15							8	15	10		7			3
12				8			17					10			11	12			19	
		17			4		10						8				14			

4

5

6

7

8

9

10

11

8		11		19	13	8	5	11				8		8		17	16		8	
13			12									10		7					12	11
12					7	21		12				14	15		8		9			
13	18	9				16								10	15				21	
			12	11			11					13			13		17	9		
	14			15		13	5	8				11		21	4				13	
	15	17			14				12			8					30			11
			24				10		18			8		8		8			7	
12					9			21			4		18	10			14			
						7		15		4	15			9						
						10		6				11	12							
						22		10		11	16			21						
10	10		7		21			17									14			13
	18					14		13			20	14		9		17				
	4	8	14		21		15							23		7			16	
14				8			16	6				10					9	5	23	8
	11	17			10							11		11		13				
12		6		17		7		14				7	10	13				16		9
	14		19	22										17			12		4	
9							8					15		6		15		15		10
	12		7		4		15					8		11						

12

5		7		15		10		15				8		22		21				
13	20			6		15						13			19	13			8	
		24					10					3		30			7		14	
17			13				15	7							9		8		11	
20			16									6		5	10		26			
19	14		13		20	9		8				11			15		3			13
		10	18				7		17			22		16		12			11	
	8				19			22	13			12				16	12			
		18							11		16		7					22		
						8		11				3	15							
							17		14	7			12							
						3		14		15	18	15								
11			13		8				3			7		16	7		10		6	10
11	17	6	9			11		12				17	11		7	11	14			
			6	13			14		21								13		23	
12	5	14		23	7			12						21	12		9		9	
					15							10			17		9			
6	9	11	15		13							17			5		13		8	
				3	13	15		12					17				11		10	11
17						11						10			13	13		21		
13		12		15			5					17								

13

14

15

16

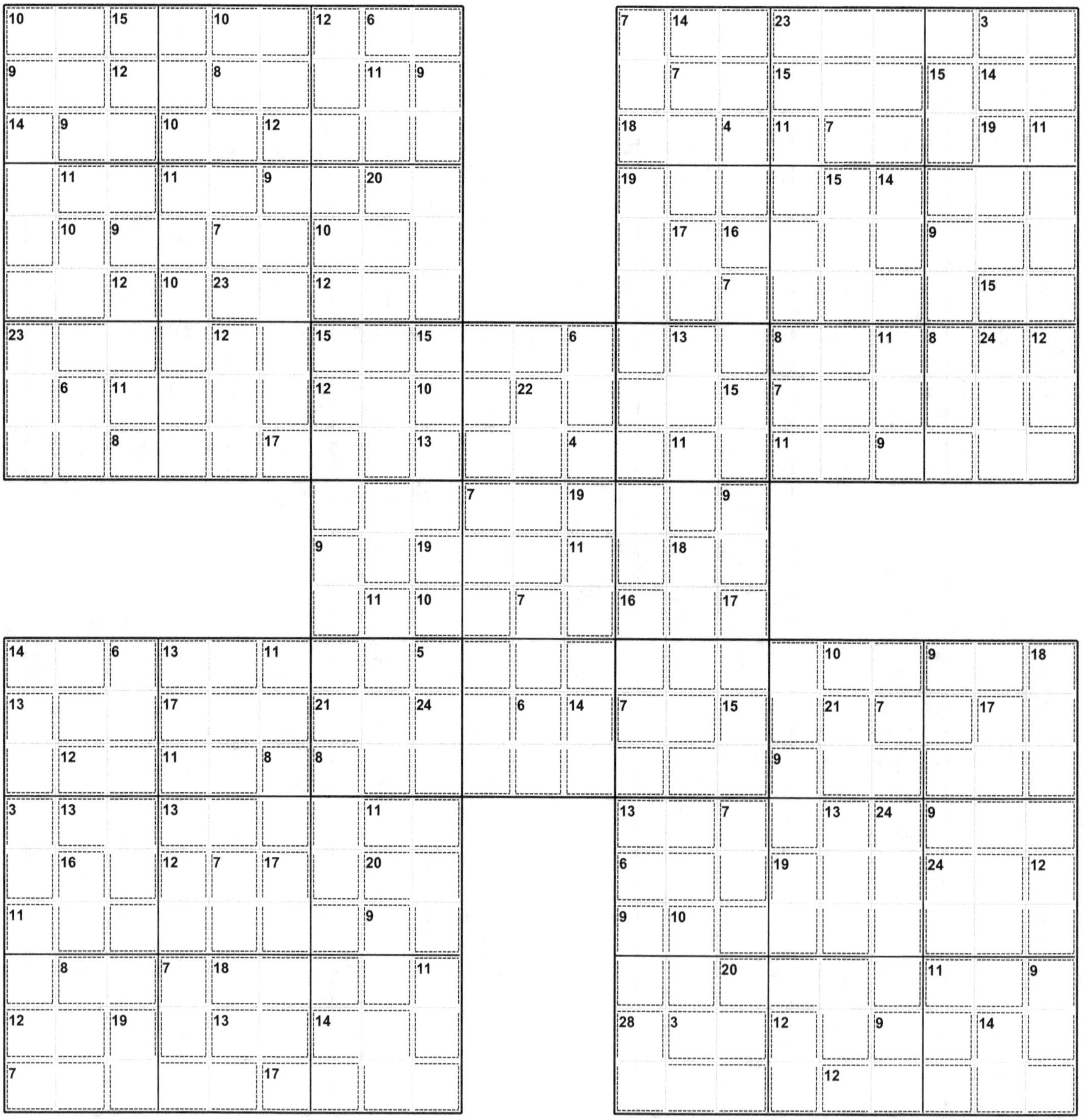

17

18

9	14	6		12	12		9	
		9	16		14		18	
13						5	9	
		16	16		12			10
13	12		17			15	10	
				9				11
16	10	11			15			
			17	12	5			7
8					17			

6		8	16	12		14		21
15	15				10		5	
		6		6		17		
11	9	10		14	24			
			3					14
14		6		14	12	11		
			10				11	
22				9		15	15	
24								

			9		15			
		7	6	13		22		
				13	13	24		
12		10					5	
16		9	18				11	
18	20		18			12	8	
			21					12
	23		17		8		21	
		13				17		

10		13						
14		17			9		23	
	8		5					13
	17	8		22		24		
13			3		10			
	12				6	8		10
11		13		13		11		
17		15			9		12	
11						17		

		12	13	11		9		24
	21			12		3		
17			15		14		9	
					5	19		3
12		8		10				
16			12		24			
11		9			15		9	22
17			18					
	8		17					

19

20

21

22

23

24

25

26

27

28

29

30

31

32

33

10	8	20	18			18						23	10		14	5		11		6
				6		21								12		15		13		
15			6		21										9		28	9		21
4		15			26		13					16		23					12	
10	7		26			12	15	9							14		9			
	11											6					5			13
12			13		5	10			8	11		23			10		15			
	13			10				12		25		17			10		15		29	
21					10		9		12				12							
						19		10				15								
								21			21									
						12		19			13		4	14						
11	14			21				10		13	14				6	13	9		15	
		10	14	14	4	11	7		18				11				12			15
16								11						27			9		3	
7		12			15	10	10					28			7	9		10		11
11	15	8						18								8			15	
				30								16				14		19		7
10	12		3		14		6					12		12					20	
	14	15		12	9	7						20			11		16			14
		9					15					11								

34

35

36

37

38

39

40

41

Top-left grid

18			18		13		23	
	20	22			16		20	
				10				
	13		9	15		17		8
12					7			
13		12		10			12	
19			19		13		11	
				14				19
10				15			20	9

Top-right grid

23				16	6	12	9	
10	11		11				10	28
	10			14				
9		23				17		
			16					
24			21	11			24	5
	12			23				
			22	3				14
	15				16			

Centre grid

	11		7		8		12	
		19			3			
	20	9		17			15	
21				10		13		
		7	16					
20				17		10	18	
	27				11			
		27	13			13		
					28			

Bottom-left grid

33				6			27	
11				24				27
20	13				8			
		8		6	13		28	
			11					
10		16		14		12	7	
9	9		9					24
		5	17	14		14		
13								

Bottom-right grid

			15		11	14		
13			8			11	11	16
				12				
		20		24	13			
9					13			10
22		11				13		
		17		5		25	14	
9			14	11			15	
10								

42

43

44

45

46

47

24			8	22			13					9		23			8		19	
10				15	15			21				19	8		8			17		
6		7											10		21					21
21			20	11	7		27						17		4		18			
					11			19						7		10	13	13		
11		9		9									9	14						8
19	16		13		6				12			20		24		12	5		8	
				13		17				11				11			19			
10				12		23				16			19		8				17	
						16	22			11				22						
							5						19							
							14	23		15		7								
23				20						16					18		14		6	
19	6		12			12		31			18		9		15		7		10	
		12		15	9	11								10		9		13		15
18			10				8					15			7		19	11		
3				17	17		17							19				17	22	
	13	15					12					19								
14		10			16			15					11	22				15		
	10		12		26									19			18			
		11										11						19		

48

15 16 7 8 22 7 13 11 11 12 20
10 17 23 10 22 20 17
14 18 10 13 13
18 22 16 14
9 23 13 13 23 16
14 12 8 17 18 3 17
14 12 6 15 15 8 18 18
4 12 11 13 21 12 13 15
8 14 22 12 17 18 6
15 18
21 15 20
11 6 17
19 20 10 27 9 24 14
15 20 15 14 15 16 18 19
17 7 12 26 10 5 24
18 9 7 22 22 10 14 18
15 14 5 5
13 12 7 10 12 19
16 26 17 11 18 14 11
20 14 8 22 16 12 7
9 11 13

49

50

51

52

53

54

55

56

57

14	10		16	14		11	10					10		8		13	6	11		9
	6						19					13			23			15		
14		17		6		15		19				10		16				7		17
		10		18								22			16		18			
11		16	9	9		14								23	5	16				15
5					9	23							28			10		6	4	
	24		17				15				21				11		9			11
11							25							11				23	16	
	19					24		15					11							
						17		11	8			25		18						
									7											
						12	15			10		12		9						
7		15					17		26		23		15		22			10	12	
13	24		7	12		10								10		7			16	7
				13										14			19			
18	18			23		10						8		11				14		18
		6					23					8		9	20					
3		12		9		8						24			16			9	9	
13		18			15		8	17				17		17			10			10
21			10											8				9		
8			7		16							12			10			21		

58

59

11			15		9	14		24				15			11	11		5	12	
14	15	5				7						9		11		19			15	
			25	25	12							13				18		17		
15					22							26			26		19			
3		19						12				10					7	13		19
12				8	15							13								
	15	19			9	11			11		17					13	7	12		
12		13				17				26	12		14						12	5
			27					17	4		22		16	31						
						21														
									21		8		9							
						11	28						9							
14	12			14					22					12		23			8	
		13		6		10			11	21	10		12			5		17		8
14		9			22			11					11		7		17	7		
16		17		22		7						18	22							16
13						10		10					13			17				
	12	5				14								21				16	20	8
10			8		5	17		5				25				5				
	15		23			7	16					6	19		15		14			
20														6					16	

60

61

62

63

64

65

12	13		15		27			8				20		10	30				9	10
	18		3		10								10		8		8			
		15		26	4		21							14		10		10		19
21		23				15						15				18		18		
6					13	8						22			12					19
							14					7		17						
	19			4	14		8	10		14			13			12	6		21	
19		13	11		16			23			6		10				10		11	
					23		16			16				16		16				
										12		8								
							13		22				20							
						7	19				15			7						
13	14			10			23	8	11			12			25				3	
	11	13		14	14				14		16	11		10		8		15		14
14		15					13						16			21				
	16		6			8		12				14		10		16	20		18	5
				18	14		11						23	11						
	19	9				17									11			14		
		20			6		11					3		18		5		20	10	18
20				15		8									22					
9						18						18					6			

66

67

68

13	11	13		5		13		
			14		8	10		15
15		14		11		13		
	16				17	13		
		14		9		26		
14			11		16			
20		6		19		12		18
19			5			7		8
						17		

9		15		14	14		10	
24			9		8	15		
				12		10		15
10		14			17			
10		18				11		7
	16				12			
	18				11		19	12
		9	15	6		20		
	6							

12		18			21		18	
7		8	13	21				9
17							6	
	9	20				18		12
		10		22			8	
17					6			14
	21	13			6	20		
			9	19			23	
12		9						

21	15			23			21	13
		15						
19			19			12		9
			12	20				
15				15		17		
22	12		10			24		
		7		15				
		10	16			14		9
13			18					

20			11		11	17		
	23		9			7		
				14		14		18
13			20				16	
6	17			13				
7					14		11	
21		13	23	11	16			15
					15		10	
11								

69

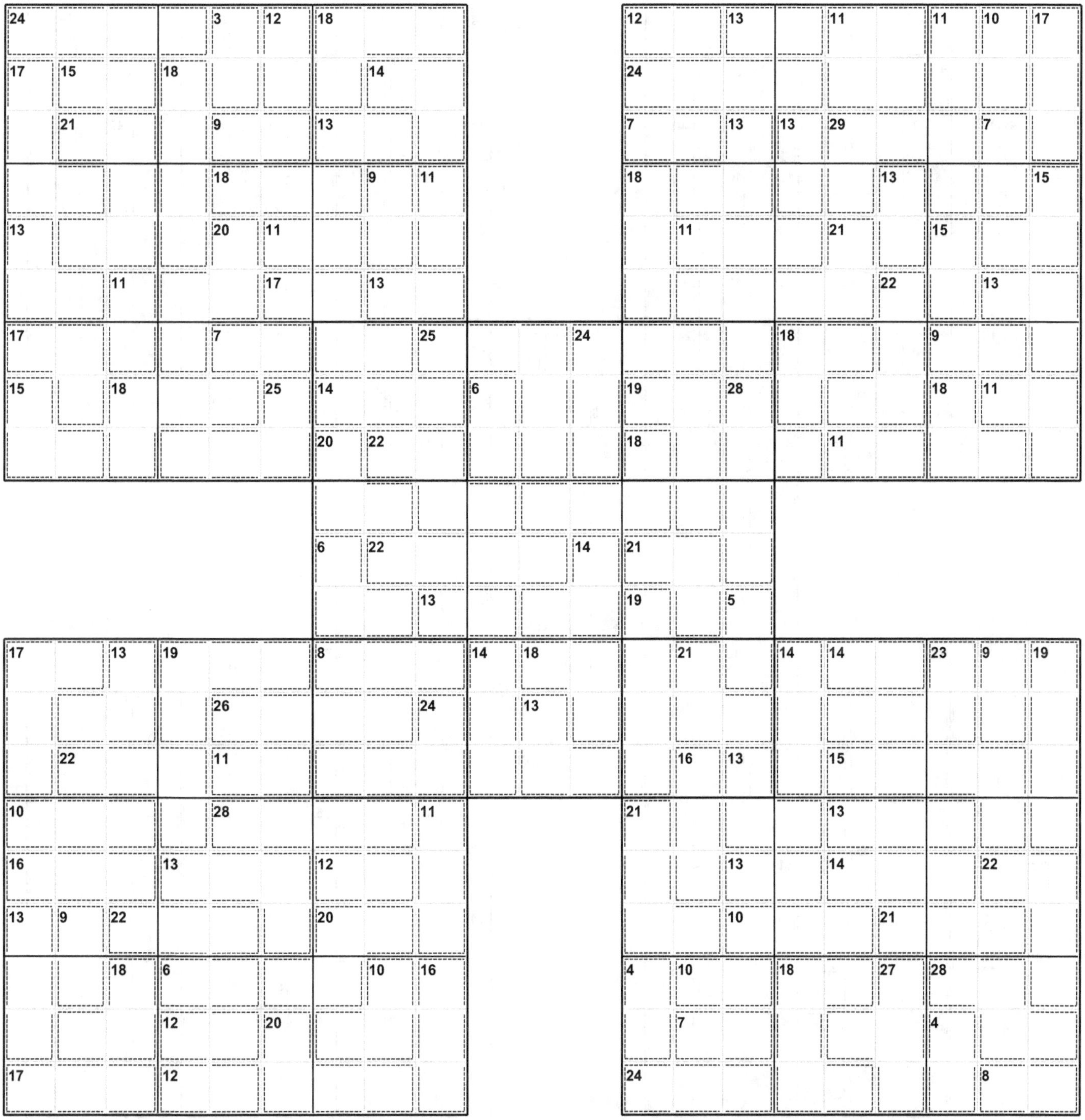

★★★★

70

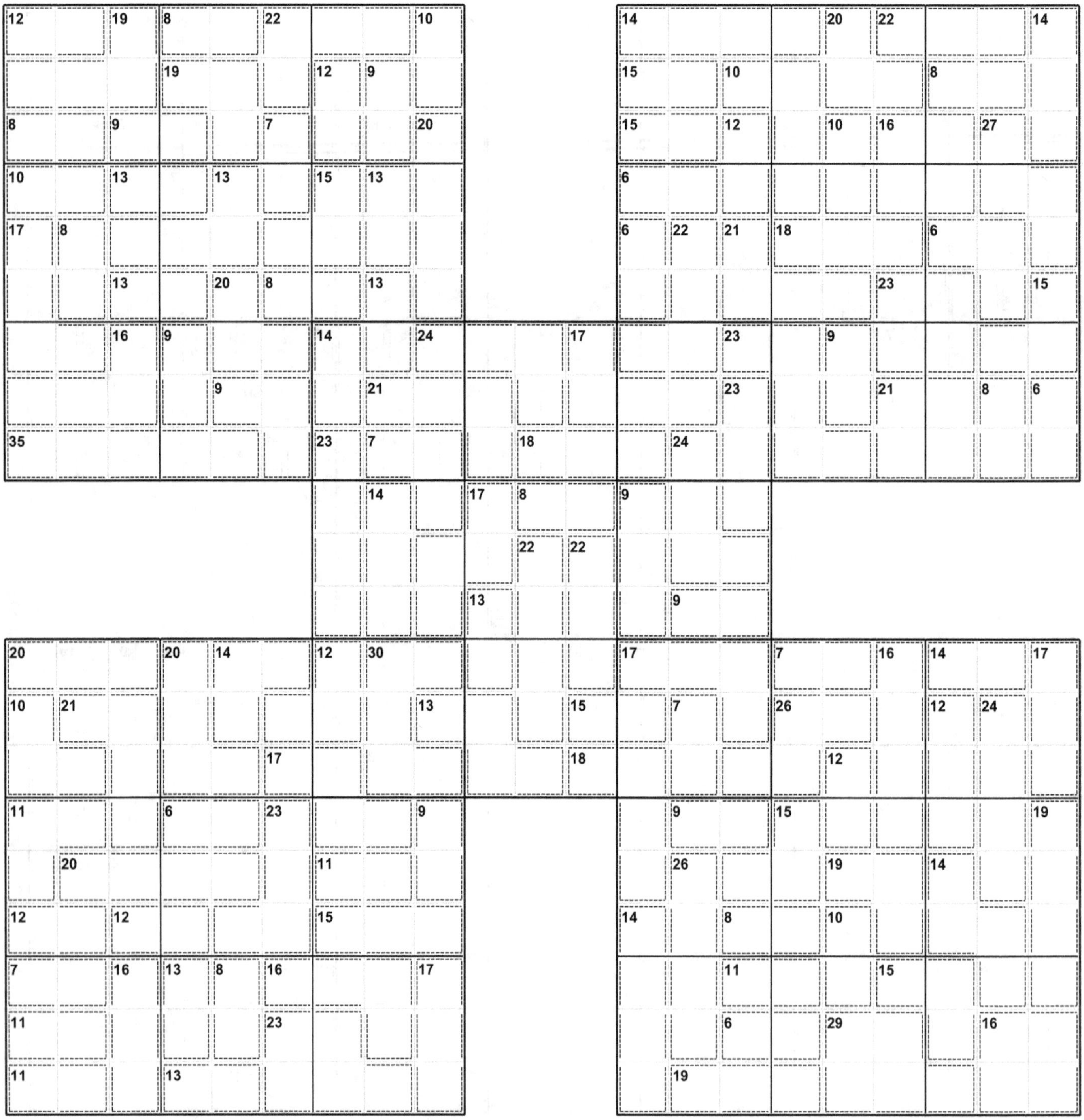
12 19 8 22 10 14 20 22 14
19 12 9 15 10 8
8 9 7 20 15 12 10 16 27
10 13 13 15 13 6
17 8 6 22 21 18 6
13 20 8 13 23 15
16 9 14 24 17 23 9
9 21 23 21 8 6
35 23 7 18 24
14 17 8 9
22 22
13 9
20 20 14 12 30 17 7 16 14 17
10 21 13 15 7 26 12 24
17 18 12
11 6 23 9 9 15 19
20 11 26 19 14
12 12 15 14 8 10
7 16 13 8 16 17 11 15
11 23 6 29 16
11 13 19

★★★★

71

72

73

74

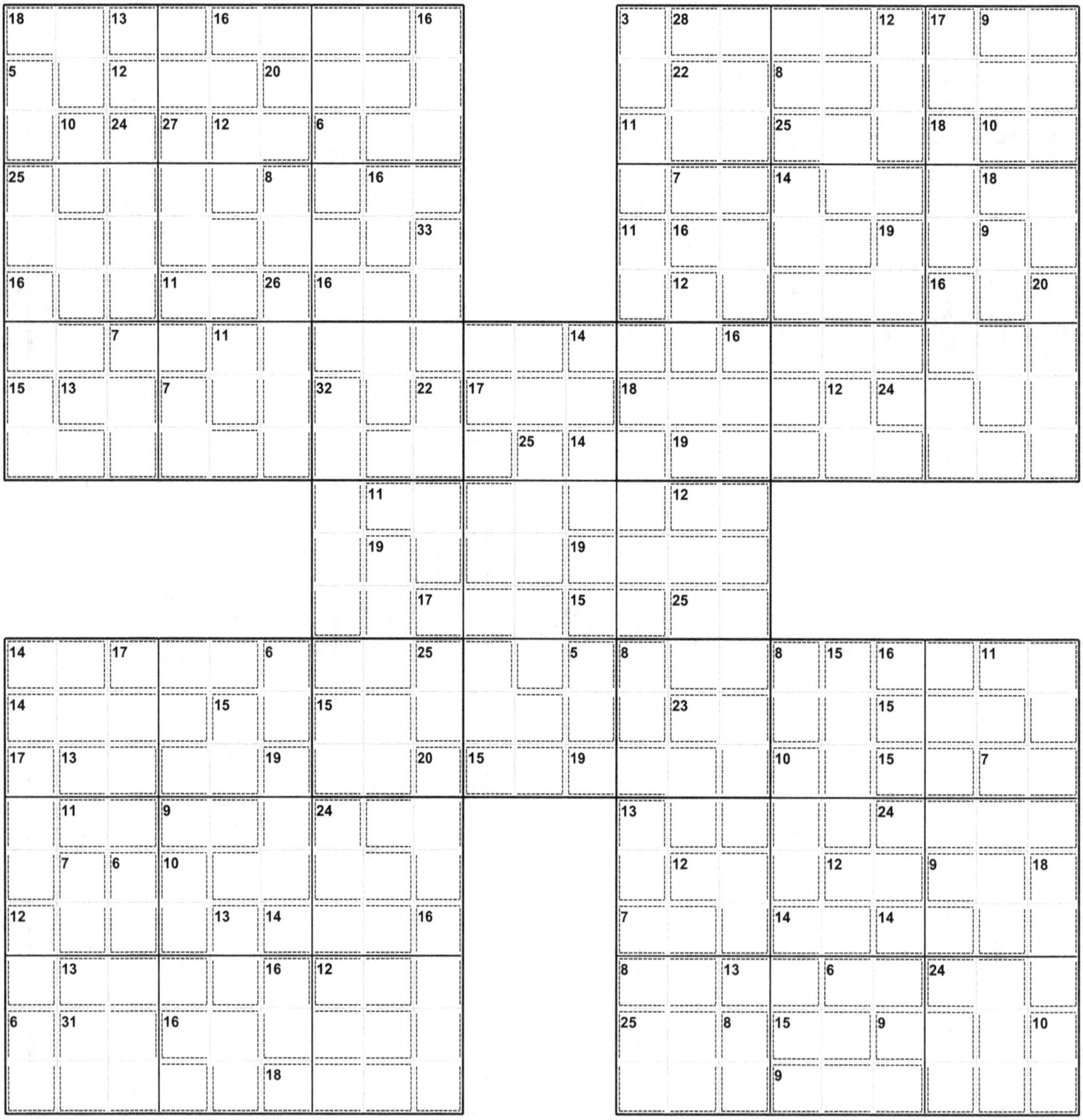

★★★★

75

76

77

78

79

80

81

82

83

84

45			31				24					9		10		5	13	45		
					17			9				17	20							
			15	21									16	25		13				
12	10	11					17							5		23		16	6	
			12	19	10			10											11	
8	6											8		11		13			18	10
		9		20		45			8	17		45			11					
20	14		9	11						14					8		7	7		13
									10						11			9		
						17	8			14		15		10						
							13	7			11		14							
								5		19										
13	17	5		15		45			13			45			19		12		9	12
		5	9						16	14						10	12	15		
10			11		14										4					16
14		11		11		11		8				19	9			9			14	
	16		9			8							9	19			12			14
5				15		19	12	12						15						
45			10		11							12				17		45		
			10				6					18			7					
			11		8		9					9		13		10				

85

86

87

88

89

90

91

92

15			16	16	17	14	13					15	19			21	23			
12							18							10			11			9
7			13				21					16				21	16	15		
14		17		16								15		16				12		10
			11				8											18		
	16		13			16						5			13		12		17	
21	12		10		14			15			20		19	18					7	
		13		9		26		24		8						24				18
				15							14	20	15							
						18			15											
						13		20						8						
							3				10	21								
10	8	10	13	26				16	9			17		19	9		18		7	
					13		8		27						6		15		12	17
20	10		10					20				17	3		14			17		
		14			17								12	12						
6	9	12		9		13						13		17		15		14		
		13		14			21								16		7	14		
	20		14			16						3				12		10	12	11
			19				22						15				25			
22												16						12		

93

94

16			22		10		15					4		17			8	20		22
16					18							13		23	22					
13			15		16			22				7			22			19		
	9	15										9				9				
32		21				22		15				16					21			15
			16		10							27		21				22		
11	10				22				17	11			9						5	
		12		12			14						25	11			25	19		
9		7						8		13				12						
						18					18			10						
						8	11		16											
							13				8		13	11						
18		11		21			25	21		14	25				11	19			12	
17			19										8			21	9		12	8
	17			11			15				17		14	12				8		
			9		19	21						7					11		19	18
		11				17							15		16					
19	13	12	10				9					10		10			15		5	
			35				10	5					21		13	6		24		9
		17			8							13							12	
7					19									17						

95

14 18 11 11 17 8 10 15 12
11 22 3 19 14 17 17 12 5
16 24 12 19 14 8
20 11 18 28 15
29 10 5 11 21
6 30 21 7 9
22 14 8 13 24 9 15 11 25
20 20 17 12 20 18 18 12
17 19 12 6
7
16 22 23 17
7 24
18 15 9 11 16 14 12 17 11 30
12 25 21 16 9 18
16 16 17 5 4 23
18 10 20 9 12 9 13
8 18 19 20
20 21 13 16 14 37
13 11 12 18 11 14 9
8 10 17 16 11
16 23 26

96

97

98

99

100

SOLUTIONS

1

6	5	2	8	4	9	1	7	3				2	8	7	1	4	5	3	6	9
8	4	7	6	1	3	9	2	5				5	4	6	9	3	2	8	1	7
1	3	9	5	2	7	8	6	4				1	3	9	8	7	6	4	2	5
7	1	5	9	8	4	2	3	6				8	9	2	6	1	3	5	7	4
2	9	4	3	6	1	5	8	7				6	1	3	4	5	7	2	9	8
3	6	8	7	5	2	4	9	1				7	5	4	2	9	8	1	3	6
4	7	1	2	3	8	6	5	9	2	8	4	3	7	1	5	8	9	6	4	2
9	8	6	4	7	5	3	1	2	5	9	7	4	6	8	7	2	1	9	5	3
5	2	3	1	9	6	7	4	8	1	3	6	9	2	5	3	6	4	7	8	1
						4	6	3	8	1	5	7	9	2						
						1	9	7	4	6	2	8	5	3						
						2	8	5	3	7	9	6	1	4						
1	7	8	4	2	9	5	3	6	7	2	8	1	4	9	8	3	5	6	7	2
5	2	9	3	1	6	8	7	4	9	5	1	2	3	6	7	9	4	1	8	5
6	3	4	8	7	5	9	2	1	6	4	3	5	8	7	6	1	2	4	3	9
8	5	2	7	6	3	4	1	9				8	9	3	2	5	6	7	4	1
9	6	3	1	4	8	2	5	7				6	2	1	4	7	9	8	5	3
4	1	7	5	9	2	6	8	3				4	7	5	1	8	3	9	2	6
2	4	5	9	3	7	1	6	8				3	6	4	9	2	7	5	1	8
3	9	6	2	8	1	7	4	5				9	5	8	3	4	1	2	6	7
7	8	1	6	5	4	3	9	2				7	1	2	5	6	8	3	9	4

2

8	2	9	4	7	5	1	6	3				4	6	8	7	5	1	9	2	3
1	3	5	9	6	2	8	7	4				5	3	9	4	2	8	6	7	1
4	7	6	3	1	8	2	9	5				2	1	7	3	9	6	8	4	5
7	1	4	8	5	9	3	2	6				3	5	2	6	8	7	1	9	4
3	9	8	6	2	7	5	4	1				8	7	6	9	1	4	3	5	2
5	6	2	1	3	4	9	8	7				9	4	1	2	3	5	7	8	6
6	4	1	2	8	3	7	5	9	1	2	3	6	8	4	1	7	2	5	3	9
2	5	3	7	9	6	4	1	8	5	9	6	7	2	3	5	6	9	4	1	8
9	8	7	5	4	1	6	3	2	8	4	7	1	9	5	8	4	3	2	6	7
						1	2	7	4	3	8	5	6	9						
						8	6	4	2	5	9	3	1	7						
						3	9	5	7	6	1	8	4	2						
1	9	3	4	2	7	5	8	6	9	7	4	2	3	1	6	7	4	8	9	5
8	7	5	1	9	6	2	4	3	6	1	5	9	7	8	1	3	5	4	6	2
2	4	6	8	5	3	9	7	1	3	8	2	4	5	6	2	8	9	1	7	3
3	1	7	2	6	9	4	5	8				3	6	4	7	2	1	5	8	9
5	8	4	3	7	1	6	2	9				1	8	5	4	9	3	6	2	7
6	2	9	5	8	4	1	3	7				7	2	9	5	6	8	3	1	4
9	5	2	6	3	8	7	1	4				8	1	2	3	5	7	9	4	6
4	6	8	7	1	5	3	9	2				5	9	7	8	4	6	2	3	1
7	3	1	9	4	2	8	6	5				6	4	3	9	1	2	7	5	8

3

4	1	6	8	2	7	3	9	5				5	4	1	6	7	3	8	2	9
9	7	8	4	3	5	1	2	6				7	2	3	5	8	9	6	1	4
2	3	5	1	6	9	7	4	8				6	9	8	4	1	2	5	3	7
8	9	1	2	7	4	6	5	3				8	5	9	2	4	1	7	6	3
5	2	7	3	8	6	4	1	9				1	7	2	3	6	8	4	9	5
6	4	3	9	5	1	8	7	2				3	6	4	7	9	5	1	8	2
1	5	2	6	4	8	9	3	7	6	2	8	4	1	5	9	3	6	2	7	8
7	8	4	5	9	3	2	6	1	5	3	4	9	8	7	1	2	4	3	5	6
3	6	9	7	1	2	5	8	4	7	9	1	2	3	6	8	5	7	9	4	1
						7	5	6	8	1	2	3	4	9						
						8	2	9	4	5	3	7	6	1						
						1	4	3	9	7	6	8	5	2						
6	8	7	1	4	2	3	9	5	1	8	7	6	2	4	5	8	7	1	9	3
3	1	9	6	8	5	4	7	2	3	6	5	1	9	8	3	4	2	5	6	7
2	5	4	9	3	7	6	1	8	2	4	9	5	7	3	1	6	9	8	2	4
4	3	6	5	1	8	9	2	7				3	4	5	6	7	8	2	1	9
9	7	5	3	2	6	8	4	1				9	6	1	2	5	3	7	4	8
8	2	1	7	9	4	5	3	6				7	8	2	9	1	4	6	3	5
1	4	2	8	6	9	7	5	3				4	5	9	8	2	6	3	7	1
7	6	3	4	5	1	2	8	9				8	3	6	7	9	1	4	5	2
5	9	8	2	7	3	1	6	4				2	1	7	4	3	5	9	8	6

4

2	3	6	7	5	9	4	1	8				3	8	4	9	2	6	7	5	1
7	5	8	2	4	1	9	3	6				7	9	1	5	4	8	2	6	3
1	4	9	6	8	3	2	7	5				2	5	6	1	7	3	8	9	4
5	9	3	8	7	4	1	6	2				4	3	2	8	5	9	6	1	7
6	1	7	9	2	5	8	4	3				5	7	9	6	1	2	3	4	8
4	8	2	3	1	6	7	5	9				6	1	8	7	3	4	5	2	9
3	2	5	1	9	7	6	8	4	5	1	3	9	2	7	3	6	1	4	8	5
9	6	1	4	3	8	5	2	7	8	4	9	1	6	3	4	8	5	9	7	2
8	7	4	5	6	2	3	9	1	2	6	7	8	4	5	2	9	7	1	3	6
						1	6	2	9	5	4	3	7	8						
						4	7	9	3	8	1	2	5	6						
						8	5	3	7	2	6	4	1	9						
4	2	6	1	8	7	9	3	5	4	7	2	6	8	1	3	2	9	7	5	4
7	5	3	6	4	9	2	1	8	6	3	5	7	9	4	6	8	5	1	2	3
8	1	9	5	3	2	7	4	6	1	9	8	5	3	2	7	1	4	8	9	6
6	7	8	9	5	3	4	2	1				9	1	7	5	3	2	6	4	8
3	4	5	8	2	1	6	7	9				3	2	8	4	6	7	5	1	9
1	9	2	7	6	4	5	8	3				4	6	5	8	9	1	2	3	7
2	6	1	3	7	5	8	9	4				8	4	3	2	5	6	9	7	1
9	8	7	4	1	6	3	5	2				2	7	9	1	4	8	3	6	5
5	3	4	2	9	8	1	6	7				1	5	6	9	7	3	4	8	2

SOLUTIONS

5

1	7	6	5	2	9	3	4	8				2	5	9	8	6	4	1	3	7
3	9	5	8	7	4	2	1	6				4	1	6	9	3	7	2	5	8
2	8	4	1	6	3	7	5	9				3	8	7	2	1	5	6	9	4
8	4	9	7	5	1	6	2	3				5	4	3	7	9	1	8	6	2
5	3	1	6	9	2	8	7	4				9	2	1	5	8	6	4	7	3
7	6	2	4	3	8	1	9	5				6	7	8	4	2	3	9	1	5
6	2	7	3	4	5	9	8	1	3	4	5	7	6	2	1	5	8	3	4	9
4	1	3	9	8	7	5	6	2	7	1	9	8	3	4	6	7	9	5	2	1
9	5	8	2	1	6	4	3	7	6	8	2	1	9	5	3	4	2	7	8	6
						6	7	9	2	5	1	3	4	8						
						8	5	3	9	6	4	2	1	7						
						1	2	4	8	3	7	9	5	6						
6	9	3	4	2	5	7	1	8	5	9	6	4	2	3	9	1	8	6	7	5
7	4	5	1	8	3	2	9	6	4	7	3	5	8	1	3	7	6	4	2	9
1	8	2	6	7	9	3	4	5	1	2	8	6	7	9	2	5	4	8	3	1
5	1	7	8	6	4	9	2	3				7	4	2	1	9	3	5	8	6
4	6	9	3	5	2	8	7	1				8	3	5	6	4	7	9	1	2
3	2	8	9	1	7	6	5	4				1	9	6	5	8	2	7	4	3
8	7	4	5	9	6	1	3	2				9	1	7	4	3	5	2	6	8
9	3	1	2	4	8	5	6	7				3	6	4	8	2	9	1	5	7
2	5	6	7	3	1	4	8	9				2	5	8	7	6	1	3	9	4

6

2	1	6	4	8	5	7	9	3				8	7	6	3	9	1	2	5	4
5	3	7	2	1	9	6	4	8				1	3	2	7	5	4	8	6	9
9	4	8	3	6	7	2	1	5				9	5	4	2	6	8	3	1	7
4	6	9	8	7	3	1	5	2				6	4	5	9	2	7	1	3	8
8	2	5	9	4	1	3	7	6				3	8	9	1	4	6	7	2	5
1	7	3	6	5	2	4	8	9				2	1	7	5	8	3	9	4	6
7	9	1	5	3	6	8	2	4	7	1	9	5	6	3	8	1	9	4	7	2
6	8	2	1	9	4	5	3	7	2	8	6	4	9	1	6	7	2	5	8	3
3	5	4	7	2	8	9	6	1	3	4	5	7	2	8	4	3	5	6	9	1
						7	9	3	5	2	1	8	4	6						
						4	8	2	6	3	7	1	5	9						
						1	5	6	8	9	4	3	7	2						
1	8	3	6	9	4	2	7	5	1	6	8	9	3	4	8	6	7	1	2	5
4	2	7	8	5	3	6	1	9	4	7	3	2	8	5	9	1	3	6	7	4
9	5	6	7	1	2	3	4	8	9	5	2	6	1	7	2	4	5	8	9	3
8	9	1	5	4	6	7	3	2				1	6	9	5	8	4	7	3	2
3	7	5	9	2	1	4	8	6				4	2	3	7	9	1	5	8	6
6	4	2	3	7	8	5	9	1				5	7	8	6	3	2	9	4	1
7	3	8	1	6	5	9	2	4				7	5	1	4	2	8	3	6	9
5	1	4	2	3	9	8	6	7				3	4	6	1	7	9	2	5	8
2	6	9	4	8	7	1	5	3				8	9	2	3	5	6	4	1	7

7

3	2	5	4	9	8	7	1	6				1	8	5	4	6	9	2	3	7
9	8	6	7	5	1	3	2	4				3	6	7	2	8	5	1	4	9
1	7	4	6	2	3	9	5	8				4	9	2	7	3	1	5	8	6
6	4	9	5	7	2	1	8	3				2	4	1	8	9	3	6	7	5
7	1	2	3	8	6	4	9	5				6	7	8	5	4	2	9	1	3
8	5	3	1	4	9	6	7	2				9	5	3	6	1	7	4	2	8
5	6	1	2	3	7	8	4	9	7	2	1	5	3	6	1	2	8	7	9	4
2	9	7	8	6	4	5	3	1	4	8	6	7	2	9	3	5	4	8	6	1
4	3	8	9	1	5	2	6	7	9	5	3	8	1	4	9	7	6	3	5	2
						6	5	4	8	1	7	2	9	3						
						9	8	3	2	6	5	1	4	7						
						1	7	2	3	9	4	6	5	8						
4	3	8	5	1	9	7	2	6	5	4	9	3	8	1	2	9	4	7	6	5
2	9	5	6	7	3	4	1	8	6	3	2	9	7	5	3	6	8	1	4	2
1	6	7	2	8	4	3	9	5	1	7	8	4	6	2	7	5	1	3	8	9
3	4	2	9	6	8	5	7	1				2	3	4	9	8	6	5	1	7
9	5	1	3	2	7	8	6	4				6	5	9	1	7	3	4	2	8
8	7	6	1	4	5	2	3	9				8	1	7	5	4	2	9	3	6
6	8	4	7	3	1	9	5	2				7	4	6	8	3	5	2	9	1
5	1	3	8	9	2	6	4	7				5	2	3	6	1	9	8	7	4
7	2	9	4	5	6	1	8	3				1	9	8	4	2	7	6	5	3

8

9	4	7	1	6	8	5	3	2				7	3	9	4	5	6	1	8	2
1	3	5	4	9	2	7	6	8				8	1	6	7	2	3	4	9	5
6	2	8	3	7	5	1	4	9				4	2	5	8	1	9	6	7	3
4	9	3	5	1	6	8	2	7				2	9	8	1	6	7	3	5	4
7	8	6	9	2	3	4	1	5				1	4	7	2	3	5	9	6	8
2	5	1	8	4	7	3	9	6				6	5	3	9	8	4	7	2	1
3	7	4	2	8	9	6	5	1	9	4	7	3	8	2	6	7	1	5	4	9
5	6	9	7	3	1	2	8	4	5	6	3	9	7	1	5	4	8	2	3	6
8	1	2	6	5	4	9	7	3	2	8	1	5	6	4	3	9	2	8	1	7
						4	6	2	7	9	5	1	3	8						
						8	1	9	3	2	4	7	5	6						
						5	3	7	8	1	6	2	4	9						
8	6	4	2	3	7	1	9	5	6	3	8	4	2	7	5	1	3	8	9	6
9	1	7	4	5	8	3	2	6	4	7	9	8	1	5	4	6	9	2	7	3
3	5	2	1	9	6	7	4	8	1	5	2	6	9	3	7	2	8	1	4	5
2	4	3	6	1	5	9	8	7				7	4	1	6	3	2	9	5	8
1	8	9	3	7	4	6	5	2				9	6	2	1	8	5	7	3	4
5	7	6	8	2	9	4	3	1				5	3	8	9	4	7	6	1	2
6	9	1	5	4	2	8	7	3				1	8	9	3	5	6	4	2	7
7	2	8	9	6	3	5	1	4				2	5	4	8	7	1	3	6	9
4	3	5	7	8	1	2	6	9				3	7	6	2	9	4	5	8	1

SOLUTIONS

9

6	9	8	1	2	5	3	7	4				7	8	2	5	1	9	4	6	3
5	1	4	6	3	7	2	8	9				6	3	9	4	7	2	5	1	8
3	7	2	4	9	8	6	1	5				4	5	1	6	3	8	9	7	2
1	3	5	2	6	9	7	4	8				2	1	8	7	9	5	6	3	4
7	4	9	5	8	3	1	2	6				9	4	3	1	2	6	8	5	7
2	8	6	7	1	4	5	9	3				5	7	6	3	8	4	1	2	9
8	6	7	3	4	1	9	5	2	8	4	1	3	6	7	8	4	1	2	9	5
4	2	1	9	5	6	8	3	7	9	5	6	1	2	4	9	5	3	7	8	6
9	5	3	8	7	2	4	6	1	2	3	7	8	9	5	2	6	7	3	4	1
						5	2	4	3	7	8	9	1	6						
						1	9	3	6	2	4	7	5	8						
						7	8	6	5	1	9	4	3	2						
2	4	6	9	5	7	3	1	8	4	6	5	2	7	9	4	1	3	6	8	5
7	1	9	3	8	6	2	4	5	7	9	3	6	8	1	2	7	5	4	9	3
5	3	8	4	1	2	6	7	9	1	8	2	5	4	3	8	6	9	2	7	1
1	2	5	7	6	3	9	8	4				1	2	5	7	4	6	9	3	8
6	8	7	1	4	9	5	2	3				3	6	7	9	2	8	5	1	4
3	9	4	5	2	8	7	6	1				8	9	4	5	3	1	7	6	2
9	6	2	8	3	4	1	5	7				9	5	2	1	8	7	3	4	6
4	5	3	6	7	1	8	9	2				7	1	6	3	5	4	8	2	9
8	7	1	2	9	5	4	3	6				4	3	8	6	9	2	1	5	7

10

1	8	7	5	2	3	4	6	9				3	2	8	9	1	5	6	7	4
9	4	2	6	7	8	1	5	3				7	5	6	2	8	4	1	3	9
3	6	5	9	1	4	8	2	7				1	9	4	3	7	6	2	5	8
2	1	6	4	9	7	5	3	8				2	4	1	5	6	8	7	9	3
7	3	4	2	8	5	6	9	1				9	6	3	4	2	7	5	8	1
8	5	9	1	3	6	7	4	2				5	8	7	1	3	9	4	2	6
4	2	3	8	6	1	9	7	5	6	3	4	8	1	2	6	5	3	9	4	7
6	9	8	7	5	2	3	1	4	2	8	5	6	7	9	8	4	2	3	1	5
5	7	1	3	4	9	2	8	6	9	1	7	4	3	5	7	9	1	8	6	2
						5	2	9	8	6	1	3	4	7						
						6	3	7	5	4	9	2	8	1						
						8	4	1	3	7	2	5	9	6						
6	8	9	1	3	7	4	5	2	7	9	8	1	6	3	9	7	8	2	4	5
2	4	7	6	8	5	1	9	3	4	2	6	7	5	8	4	2	6	1	3	9
5	3	1	2	4	9	7	6	8	1	5	3	9	2	4	3	5	1	8	7	6
9	7	4	5	1	2	3	8	6				6	1	7	8	3	2	5	9	4
8	6	2	3	9	4	5	7	1				3	9	2	6	4	5	7	8	1
3	1	5	7	6	8	9	2	4				4	8	5	1	9	7	6	2	3
7	5	3	4	2	6	8	1	9				8	3	1	2	6	9	4	5	7
1	2	8	9	7	3	6	4	5				2	7	9	5	1	4	3	6	8
4	9	6	8	5	1	2	3	7				5	4	6	7	8	3	9	1	2

11

7	1	8	3	6	9	5	4	2				5	3	1	7	9	4	8	2	6
6	5	2	7	8	4	3	1	9				4	6	2	5	8	3	1	9	7
9	3	4	1	5	2	6	7	8				9	7	8	2	6	1	5	3	4
3	9	1	6	2	5	7	8	4				1	4	7	8	5	2	3	6	9
4	2	7	8	1	3	9	5	6				8	5	3	9	4	6	2	7	1
5	8	6	4	9	7	1	2	3				2	9	6	1	3	7	4	8	5
1	7	9	2	4	6	8	3	5	6	4	2	7	1	9	4	2	8	6	5	3
2	6	3	5	7	8	4	9	1	3	8	7	6	2	5	3	1	9	7	4	8
8	4	5	9	3	1	2	6	7	5	9	1	3	8	4	6	7	5	9	1	2
						5	2	8	7	3	9	4	6	1						
						3	7	4	2	1	6	9	5	8						
						9	1	6	4	5	8	2	7	3						
1	2	8	3	4	6	7	5	9	8	6	3	1	4	2	9	7	3	6	5	8
4	9	7	2	1	5	6	8	3	1	2	4	5	9	7	2	6	8	3	4	1
5	3	6	8	9	7	1	4	2	9	7	5	8	3	6	5	1	4	2	7	9
8	1	2	6	3	4	9	7	5				2	7	1	3	9	6	4	8	5
6	4	9	7	5	2	8	3	1				6	5	4	7	8	2	1	9	3
3	7	5	1	8	9	2	6	4				3	8	9	4	5	1	7	6	2
9	6	1	4	7	3	5	2	8				4	2	8	6	3	5	9	1	7
2	5	3	9	6	8	4	1	7				9	6	5	1	2	7	8	3	4
7	8	4	5	2	1	3	9	6				7	1	3	8	4	9	5	2	6

12

3	2	6	1	8	7	4	5	9				5	3	8	7	1	9	4	6	2
5	9	8	3	2	4	7	1	6				4	9	6	5	2	3	8	1	7
1	7	4	5	6	9	3	2	8				2	1	7	8	6	4	3	9	5
9	1	7	2	3	8	5	6	4				6	8	9	4	5	1	7	2	3
8	4	5	6	7	1	2	9	3				1	5	2	3	7	8	9	4	6
2	6	3	9	4	5	1	8	7				7	4	3	6	9	2	1	5	8
7	8	2	4	5	6	9	3	1	6	4	2	8	7	5	9	4	6	2	3	1
6	5	1	7	9	3	8	4	2	7	3	5	9	6	1	2	3	7	5	8	4
4	3	9	8	1	2	6	7	5	8	1	9	3	2	4	1	8	5	6	7	9
						5	8	4	3	2	7	1	9	6						
						3	9	7	5	6	1	2	4	8						
						2	1	6	9	8	4	7	5	3						
6	3	2	9	4	7	1	5	8	2	7	6	4	3	9	6	1	8	2	5	7
7	8	5	6	3	1	4	2	9	1	5	3	6	8	7	4	2	5	9	1	3
4	9	1	2	5	8	7	6	3	4	9	8	5	1	2	3	9	7	6	4	8
3	1	7	4	9	2	5	8	6				2	4	6	7	5	1	8	3	9
9	4	6	1	8	5	3	7	2				3	7	1	9	8	4	5	6	2
5	2	8	7	6	3	9	1	4				8	9	5	2	3	6	4	7	1
1	7	3	8	2	4	6	9	5				9	5	4	1	7	2	3	8	6
2	6	4	5	1	9	8	3	7				1	6	3	8	4	9	7	2	5
8	5	9	3	7	6	2	4	1				7	2	8	5	6	3	1	9	4

SOLUTIONS

13

6	8	7	4	2	5	3	9	1				4	7	8	3	1	2	9	5	6
9	2	5	7	1	3	6	4	8				6	3	2	5	9	7	8	1	4
1	4	3	6	8	9	7	2	5				1	9	5	6	8	4	2	7	3
7	9	1	5	4	6	2	8	3				7	4	9	8	3	5	6	2	1
5	6	4	2	3	8	1	7	9				2	6	3	4	7	1	5	8	9
2	3	8	9	7	1	5	6	4				8	5	1	2	6	9	3	4	7
3	5	2	8	6	4	9	1	7	3	2	4	5	8	6	7	4	3	1	9	2
4	7	9	1	5	2	8	3	6	5	7	1	9	2	4	1	5	6	7	3	8
8	1	6	3	9	7	4	5	2	8	9	6	3	1	7	9	2	8	4	6	5
						3	9	8	2	4	5	7	6	1						
						5	7	1	6	3	8	4	9	2						
						2	6	4	7	1	9	8	5	3						
4	8	1	5	3	6	7	2	9	1	8	3	6	4	5	2	7	3	1	8	9
6	3	7	2	8	9	1	4	5	9	6	7	2	3	8	9	6	1	5	4	7
2	9	5	4	1	7	6	8	3	4	5	2	1	7	9	5	8	4	6	2	3
9	4	8	1	7	3	5	6	2				5	8	3	4	9	7	2	1	6
3	5	6	9	2	4	8	7	1				7	6	1	3	2	5	8	9	4
1	7	2	8	6	5	9	3	4				4	9	2	8	1	6	7	3	5
8	1	3	7	9	2	4	5	6				9	5	7	1	3	8	4	6	2
5	2	9	6	4	8	3	1	7				8	2	6	7	4	9	3	5	1
7	6	4	3	5	1	2	9	8				3	1	4	6	5	2	9	7	8

14

9	5	4	8	3	2	7	1	6				9	4	3	8	7	5	2	6	1
3	6	7	9	1	5	4	2	8				2	1	7	9	3	6	5	4	8
8	2	1	4	7	6	3	5	9				8	6	5	4	2	1	7	3	9
7	4	9	5	8	3	1	6	2				6	8	9	7	4	3	1	2	5
2	3	8	1	6	9	5	4	7				7	2	1	6	5	9	4	8	3
6	1	5	2	4	7	8	9	3				5	3	4	2	1	8	6	9	7
1	8	2	7	9	4	6	3	5	4	7	2	1	9	8	5	6	2	3	7	4
5	7	6	3	2	1	9	8	4	6	1	5	3	7	2	1	9	4	8	5	6
4	9	3	6	5	8	2	7	1	3	9	8	4	5	6	3	8	7	9	1	2
						8	6	2	5	3	9	7	4	1						
						5	1	7	2	6	4	9	8	3						
						4	9	3	1	8	7	2	6	5						
2	3	1	6	9	4	7	5	8	9	2	3	6	1	4	8	9	3	5	2	7
8	9	7	1	5	2	3	4	6	7	5	1	8	2	9	7	5	4	3	1	6
5	6	4	8	7	3	1	2	9	8	4	6	5	3	7	6	1	2	4	8	9
7	1	6	3	2	9	4	8	5				7	6	5	4	8	9	2	3	1
4	2	5	7	8	1	9	6	3				1	9	2	5	3	6	8	7	4
9	8	3	5	4	6	2	1	7				3	4	8	1	2	7	6	9	5
3	5	2	9	1	8	6	7	4				9	8	6	2	7	5	1	4	3
6	4	8	2	3	7	5	9	1				2	5	3	9	4	1	7	6	8
1	7	9	4	6	5	8	3	2				4	7	1	3	6	8	9	5	2

15

5	6	8	1	7	3	9	4	2				5	3	7	2	1	4	9	8	6
7	2	4	9	8	6	5	3	1				6	8	9	7	5	3	4	2	1
9	3	1	2	5	4	6	7	8				4	2	1	8	6	9	3	5	7
3	4	6	8	1	5	2	9	7				9	4	8	1	2	5	7	6	3
8	7	2	4	6	9	3	1	5				1	7	3	6	9	8	5	4	2
1	9	5	3	2	7	4	8	6				2	6	5	4	3	7	8	1	9
6	8	9	7	3	2	1	5	4	7	2	8	3	9	6	5	4	1	2	7	8
2	1	3	5	4	8	7	6	9	1	4	3	8	5	2	9	7	6	1	3	4
4	5	7	6	9	1	8	2	3	5	6	9	7	1	4	3	8	2	6	9	5
						9	1	5	6	7	2	4	8	3						
						2	3	6	4	8	5	9	7	1						
						4	7	8	3	9	1	2	6	5						
3	5	8	7	4	1	6	9	2	8	1	4	5	3	7	8	6	9	4	2	1
4	7	2	5	9	6	3	8	1	2	5	7	6	4	9	1	3	2	5	8	7
9	6	1	3	2	8	5	4	7	9	3	6	1	2	8	4	5	7	3	6	9
5	8	6	4	3	7	1	2	9				7	5	4	9	2	8	6	1	3
7	3	9	2	1	5	4	6	8				8	1	3	5	4	6	7	9	2
1	2	4	8	6	9	7	5	3				9	6	2	7	1	3	8	5	4
6	4	3	1	8	2	9	7	5				2	7	5	6	9	4	1	3	8
8	9	5	6	7	3	2	1	4				4	9	1	3	8	5	2	7	6
2	1	7	9	5	4	8	3	6				3	8	6	2	7	1	9	4	5

16

3	7	6	9	2	8	4	5	1				5	6	8	9	7	3	4	2	1
4	5	9	3	7	1	8	2	6				2	3	4	8	1	6	7	5	9
2	8	1	4	6	5	7	9	3				9	7	1	4	5	2	8	6	3
5	9	2	7	4	6	3	1	8				1	2	3	7	8	5	9	4	6
7	3	8	1	5	2	6	4	9				6	8	7	3	4	9	5	1	2
1	6	4	8	9	3	5	7	2				4	9	5	6	2	1	3	8	7
8	1	7	2	3	4	9	6	5	7	3	4	8	1	2	5	3	7	6	9	4
9	2	5	6	8	7	1	3	4	6	8	2	7	5	9	1	6	4	2	3	8
6	4	3	5	1	9	2	8	7	5	9	1	3	4	6	2	9	8	1	7	5
						6	1	3	2	5	9	4	7	8						
						5	2	8	4	7	3	6	9	1						
						4	7	9	1	6	8	2	3	5						
9	5	1	7	6	3	8	4	2	3	1	5	9	6	7	1	8	2	4	5	3
6	3	2	4	8	5	7	9	1	8	4	6	5	2	3	4	7	6	1	8	9
7	4	8	9	2	1	3	5	6	9	2	7	1	8	4	3	5	9	2	7	6
1	6	3	8	5	7	4	2	9				8	5	1	6	9	7	3	2	4
2	7	4	6	3	9	1	8	5				2	4	6	8	3	5	9	1	7
8	9	5	1	4	2	6	3	7				3	7	9	2	1	4	8	6	5
3	1	7	2	9	4	5	6	8				6	3	5	9	2	8	7	4	1
4	8	9	5	7	6	2	1	3				7	1	2	5	4	3	6	9	8
5	2	6	3	1	8	9	7	4				4	9	8	7	6	1	5	3	2

SOLUTIONS

17

4	2	6	3	5	8	1	7	9				3	5	8	4	2	9	7	1	6
7	5	9	6	2	1	3	4	8				9	6	1	3	8	7	4	5	2
8	1	3	9	4	7	2	6	5				4	2	7	1	5	6	8	9	3
9	4	8	1	7	2	5	3	6				7	8	4	6	9	1	2	3	5
5	3	7	4	6	9	8	1	2				1	9	5	2	3	4	6	7	8
2	6	1	5	8	3	7	9	4				2	3	6	5	7	8	1	4	9
3	9	4	2	1	5	6	8	7	3	9	4	5	1	2	7	6	3	9	8	4
6	7	5	8	3	4	9	2	1	5	6	7	8	4	3	9	1	2	5	6	7
1	8	2	7	9	6	4	5	3	2	1	8	6	7	9	8	4	5	3	2	1
						5	1	2	4	8	6	9	3	7						
						3	9	8	1	7	2	4	5	6						
						7	4	6	9	5	3	1	2	8						
1	3	6	2	5	9	8	7	4	6	3	5	2	9	1	5	7	3	4	6	8
4	2	9	8	6	7	1	3	5	8	2	9	7	6	4	8	1	2	9	3	5
8	7	5	3	1	4	2	6	9	7	4	1	3	8	5	4	6	9	7	2	1
6	8	7	9	3	2	5	4	1				8	3	9	1	4	7	6	5	2
9	1	2	6	4	5	3	8	7				1	7	2	6	9	5	8	4	3
3	5	4	1	7	8	9	2	6				4	5	6	2	3	8	1	7	9
2	6	1	4	9	3	7	5	8				9	4	8	3	2	6	5	1	7
7	4	3	5	8	1	6	9	2				5	1	3	7	8	4	2	9	6
5	9	8	7	2	6	4	1	3				6	2	7	9	5	1	3	8	4

18

6	9	4	2	3	7	5	8	1				2	4	5	7	3	9	8	6	1
3	5	1	4	9	8	6	2	7				9	7	3	8	1	6	4	2	5
2	7	8	5	1	6	3	4	9				6	8	1	5	4	2	7	3	9
1	3	7	6	8	9	2	5	4				8	2	7	3	5	4	1	9	6
5	4	9	7	2	3	8	1	6				3	1	6	2	9	7	5	8	4
8	6	2	1	4	5	7	9	3				5	9	4	1	6	8	2	7	3
9	2	5	3	6	1	4	7	8	6	3	9	1	5	2	6	8	3	9	4	7
7	8	6	9	5	4	1	3	2	4	5	8	7	6	9	4	2	1	3	5	8
4	1	3	8	7	2	9	6	5	2	7	1	4	3	8	9	7	5	6	1	2
						8	4	1	9	6	7	5	2	3						
						7	9	6	5	2	3	8	4	1						
						2	5	3	1	8	4	9	7	6						
9	1	8	5	7	3	6	2	4	8	9	5	3	1	7	9	6	2	5	4	8
4	5	7	8	2	6	3	1	9	7	4	2	6	8	5	4	7	3	2	1	9
3	2	6	4	1	9	5	8	7	3	1	6	2	9	4	1	5	8	6	3	7
2	8	1	7	3	4	9	5	6				4	3	8	5	9	1	7	6	2
6	4	5	1	9	8	2	7	3				5	7	2	6	8	4	3	9	1
7	3	9	2	6	5	1	4	8				1	6	9	3	2	7	8	5	4
5	6	4	9	8	1	7	3	2				7	4	3	8	1	6	9	2	5
8	9	2	3	5	7	4	6	1				8	5	1	2	3	9	4	7	6
1	7	3	6	4	2	8	9	5				9	2	6	7	4	5	1	8	3

19

2	4	1	6	7	8	3	9	5				5	4	6	3	2	1	7	9	8
8	6	3	2	5	9	7	4	1				9	7	2	8	6	4	3	1	5
7	5	9	3	4	1	8	6	2				8	3	1	7	5	9	6	2	4
3	7	6	5	1	4	9	2	8				7	6	4	1	8	2	5	3	9
5	2	8	7	9	3	4	1	6				1	2	9	5	3	6	8	4	7
9	1	4	8	2	6	5	3	7				3	5	8	9	4	7	2	6	1
4	3	5	1	6	7	2	8	9	4	5	7	6	1	3	4	7	5	9	8	2
1	9	7	4	8	2	6	5	3	1	2	9	4	8	7	2	9	3	1	5	6
6	8	2	9	3	5	1	7	4	3	6	8	2	9	5	6	1	8	4	7	3
						3	1	5	2	7	6	8	4	9						
						4	2	8	5	9	3	1	7	6						
						9	6	7	8	1	4	3	5	2						
2	9	5	7	1	3	8	4	6	7	3	5	9	2	1	7	3	4	5	8	6
1	7	6	4	9	8	5	3	2	9	8	1	7	6	4	5	1	8	3	2	9
3	4	8	5	2	6	7	9	1	6	4	2	5	3	8	9	2	6	1	7	4
6	1	7	2	8	4	3	5	9				8	7	6	1	9	2	4	5	3
8	2	9	3	5	1	4	6	7				1	9	3	8	4	5	7	6	2
5	3	4	6	7	9	2	1	8				2	4	5	6	7	3	8	9	1
9	5	1	8	4	7	6	2	3				4	8	7	2	6	1	9	3	5
7	6	2	1	3	5	9	8	4				3	5	2	4	8	9	6	1	7
4	8	3	9	6	2	1	7	5				6	1	9	3	5	7	2	4	8

20

3	2	5	8	7	9	6	1	4				7	2	3	9	8	1	5	4	6
1	7	4	5	2	6	8	9	3				6	5	1	2	7	4	8	3	9
8	9	6	3	1	4	2	5	7				9	4	8	3	6	5	1	2	7
9	3	8	2	4	1	5	7	6				5	7	6	8	4	9	2	1	3
4	1	7	6	5	8	9	3	2				4	3	9	7	1	2	6	5	8
6	5	2	7	9	3	1	4	8				8	1	2	6	5	3	7	9	4
7	6	3	1	8	5	4	2	9	8	1	7	3	6	5	4	2	8	9	7	1
5	8	9	4	3	2	7	6	1	3	5	9	2	8	4	1	9	7	3	6	5
2	4	1	9	6	7	3	8	5	6	4	2	1	9	7	5	3	6	4	8	2
						8	5	6	7	3	1	9	4	2						
						9	4	3	2	6	5	7	1	8						
						2	1	7	4	9	8	6	5	3						
3	2	4	6	7	5	1	9	8	5	2	3	4	7	6	3	5	8	2	1	9
7	6	1	9	2	8	5	3	4	1	7	6	8	2	9	4	1	7	5	3	6
9	8	5	3	4	1	6	7	2	9	8	4	5	3	1	9	6	2	7	8	4
2	3	6	7	8	9	4	1	5				1	6	3	8	4	5	9	2	7
5	1	7	2	3	4	9	8	6				9	5	8	7	2	1	6	4	3
8	4	9	5	1	6	7	2	3				2	4	7	6	9	3	8	5	1
4	7	2	1	6	3	8	5	9				6	1	2	5	7	4	3	9	8
1	5	8	4	9	2	3	6	7				7	8	4	2	3	9	1	6	5
6	9	3	8	5	7	2	4	1				3	9	5	1	8	6	4	7	2

SOLUTIONS

21

5	8	3	6	9	7	2	1	4				7	6	8	1	3	2	4	9	5
1	9	4	8	5	2	6	3	7				4	3	5	8	9	6	1	2	7
6	2	7	1	4	3	8	9	5				9	1	2	7	4	5	6	8	3
4	7	5	9	2	8	1	6	3				3	2	1	4	6	7	9	5	8
8	6	1	7	3	5	4	2	9				6	5	9	3	8	1	2	7	4
9	3	2	4	6	1	7	5	8				8	4	7	2	5	9	3	6	1
7	1	9	3	8	6	5	4	2	3	7	8	1	9	6	5	7	4	8	3	2
2	4	8	5	1	9	3	7	6	2	1	9	5	8	4	6	2	3	7	1	9
3	5	6	2	7	4	9	8	1	6	5	4	2	7	3	9	1	8	5	4	6
						1	3	4	7	8	6	9	5	2						
						7	6	9	5	2	3	4	1	8						
						2	5	8	9	4	1	6	3	7						
4	5	9	7	1	8	6	2	3	1	9	7	8	4	5	3	7	1	2	9	6
3	8	6	2	9	5	4	1	7	8	6	5	3	2	9	6	4	8	5	1	7
7	2	1	4	6	3	8	9	5	4	3	2	7	6	1	2	9	5	8	4	3
1	3	8	6	4	7	9	5	2				6	9	7	8	1	4	3	5	2
2	9	5	3	8	1	7	6	4				5	8	2	9	6	3	1	7	4
6	4	7	5	2	9	1	3	8				1	3	4	7	5	2	9	6	8
9	1	4	8	5	2	3	7	6				2	7	6	1	8	9	4	3	5
8	7	2	1	3	6	5	4	9				9	5	3	4	2	6	7	8	1
5	6	3	9	7	4	2	8	1				4	1	8	5	3	7	6	2	9

22

9	7	8	3	1	5	6	4	2				7	8	5	6	9	3	1	4	2
4	5	6	7	2	9	1	3	8				3	6	1	2	8	4	9	5	7
3	1	2	6	4	8	9	7	5				9	4	2	5	7	1	8	6	3
8	4	7	9	3	2	5	6	1				1	7	3	9	5	2	4	8	6
1	2	3	5	8	6	7	9	4				6	5	9	8	4	7	2	3	1
5	6	9	4	7	1	8	2	3				4	2	8	3	1	6	5	7	9
2	9	4	1	5	7	3	8	6	9	4	5	2	1	7	4	3	5	6	9	8
6	8	5	2	9	3	4	1	7	2	3	8	5	9	6	7	2	8	3	1	4
7	3	1	8	6	4	2	5	9	6	7	1	8	3	4	1	6	9	7	2	5
						9	4	2	3	6	7	1	8	5						
						8	7	3	5	1	4	9	6	2						
						1	6	5	8	2	9	4	7	3						
2	6	4	7	1	9	5	3	8	7	9	2	6	4	1	9	3	2	8	7	5
9	5	7	4	3	8	6	2	1	4	8	3	7	5	9	1	4	8	3	6	2
8	1	3	6	2	5	7	9	4	1	5	6	3	2	8	5	7	6	1	4	9
3	2	6	8	9	1	4	5	7				2	6	4	3	5	1	7	9	8
5	4	8	3	7	6	9	1	2				1	7	5	8	2	9	4	3	6
1	7	9	2	5	4	8	6	3				8	9	3	7	6	4	2	5	1
7	8	1	9	6	2	3	4	5				5	3	2	6	1	7	9	8	4
4	9	5	1	8	3	2	7	6				9	1	7	4	8	5	6	2	3
6	3	2	5	4	7	1	8	9				4	8	6	2	9	3	5	1	7

23

3	5	1	6	9	8	2	4	7				8	7	5	6	3	1	2	4	9
6	7	9	4	1	2	3	5	8				4	3	9	5	8	2	1	7	6
2	8	4	7	3	5	9	6	1				2	6	1	7	9	4	5	8	3
8	6	3	2	4	7	5	1	9				9	2	8	1	7	6	4	3	5
7	9	2	8	5	1	4	3	6				1	4	7	8	5	3	6	9	2
1	4	5	3	6	9	7	8	2				3	5	6	4	2	9	8	1	7
9	1	8	5	7	4	6	2	3	7	1	9	5	8	4	9	6	7	3	2	1
5	3	7	1	2	6	8	9	4	5	3	6	7	1	2	3	4	5	9	6	8
4	2	6	9	8	3	1	7	5	2	8	4	6	9	3	2	1	8	7	5	4
						9	8	6	4	2	1	3	7	5						
						3	4	7	6	5	8	1	2	9						
						5	1	2	3	9	7	4	6	8						
4	5	8	7	9	3	2	6	1	8	4	5	9	3	7	2	5	6	8	4	1
9	3	1	6	4	2	7	5	8	9	6	3	2	4	1	8	7	9	5	6	3
6	7	2	8	5	1	4	3	9	1	7	2	8	5	6	3	4	1	7	9	2
2	9	5	3	8	7	1	4	6				3	9	4	5	2	7	1	8	6
8	6	7	5	1	4	3	9	2				1	7	5	9	6	8	3	2	4
1	4	3	9	2	6	8	7	5				6	8	2	4	1	3	9	5	7
3	1	6	2	7	5	9	8	4				5	6	8	1	3	4	2	7	9
5	8	4	1	3	9	6	2	7				4	1	9	7	8	2	6	3	5
7	2	9	4	6	8	5	1	3				7	2	3	6	9	5	4	1	8

24

6	2	5	7	4	9	1	8	3				1	4	6	9	2	3	5	7	8
9	1	8	6	5	3	4	7	2				9	5	8	7	6	4	1	2	3
4	3	7	1	2	8	5	9	6				3	2	7	8	5	1	6	4	9
8	4	9	5	3	6	7	2	1				7	6	4	1	8	9	3	5	2
7	5	3	2	8	1	9	6	4				2	1	5	3	7	6	9	8	4
1	6	2	4	9	7	3	5	8				8	9	3	5	4	2	7	6	1
2	7	1	3	6	5	8	4	9	3	2	5	6	7	1	4	3	8	2	9	5
3	8	4	9	7	2	6	1	5	9	8	7	4	3	2	6	9	5	8	1	7
5	9	6	8	1	4	2	3	7	4	6	1	5	8	9	2	1	7	4	3	6
						5	6	8	1	4	2	3	9	7						
						1	7	4	5	9	3	8	2	6						
						9	2	3	8	7	6	1	4	5						
9	2	8	3	7	1	4	5	6	2	3	9	7	1	8	3	4	6	5	9	2
6	3	1	5	4	9	7	8	2	6	1	4	9	5	3	8	1	2	6	4	7
4	5	7	6	2	8	3	9	1	7	5	8	2	6	4	5	7	9	3	8	1
5	1	4	9	6	2	8	7	3				1	7	5	9	8	4	2	6	3
7	6	2	8	3	4	9	1	5				3	8	2	6	5	1	4	7	9
8	9	3	1	5	7	2	6	4				6	4	9	2	3	7	8	1	5
3	7	5	4	8	6	1	2	9				8	2	1	4	9	3	7	5	6
1	8	6	2	9	3	5	4	7				5	9	6	7	2	8	1	3	4
2	4	9	7	1	5	6	3	8				4	3	7	1	6	5	9	2	8

SOLUTIONS

25

9	1	8	3	2	7	5	4	6				7	5	1	9	2	8	4	6	3
3	4	6	9	8	5	7	1	2				8	2	3	6	7	4	9	5	1
5	2	7	4	6	1	8	3	9				4	6	9	5	3	1	7	8	2
6	7	4	5	9	8	3	2	1				9	4	5	3	1	6	2	7	8
8	3	2	7	1	6	4	9	5				1	7	8	4	9	2	6	3	5
1	5	9	2	3	4	6	7	8				2	3	6	8	5	7	1	9	4
7	8	3	1	5	9	2	6	4	1	3	9	5	8	7	1	4	9	3	2	6
4	9	5	6	7	2	1	8	3	2	5	7	6	9	4	2	8	3	5	1	7
2	6	1	8	4	3	9	5	7	4	6	8	3	1	2	7	6	5	8	4	9
						5	9	2	7	1	4	8	3	6						
						6	3	1	8	2	5	4	7	9						
						4	7	8	3	9	6	1	2	5						
8	5	3	9	4	2	7	1	6	5	8	2	9	4	3	8	5	6	7	2	1
7	6	2	1	5	8	3	4	9	6	7	1	2	5	8	7	1	4	3	6	9
4	9	1	6	3	7	8	2	5	9	4	3	7	6	1	9	3	2	8	4	5
9	4	7	8	6	5	1	3	2				5	2	6	4	8	7	1	9	3
5	1	6	7	2	3	9	8	4				1	7	9	6	2	3	4	5	8
2	3	8	4	1	9	6	5	7				8	3	4	1	9	5	2	7	6
6	2	9	5	8	1	4	7	3				6	1	7	3	4	9	5	8	2
3	8	4	2	7	6	5	9	1				4	8	5	2	6	1	9	3	7
1	7	5	3	9	4	2	6	8				3	9	2	5	7	8	6	1	4

26

4	6	5	9	3	7	2	8	1				4	9	3	1	5	6	8	7	2
3	7	8	1	4	2	6	5	9				8	6	7	3	2	4	1	9	5
9	2	1	6	8	5	7	3	4				5	2	1	7	9	8	6	3	4
8	3	7	5	2	4	1	9	6				2	8	5	9	3	7	4	6	1
1	4	6	7	9	3	8	2	5				6	7	9	4	1	2	5	8	3
2	5	9	8	1	6	3	4	7				3	1	4	8	6	5	7	2	9
7	8	2	4	6	9	5	1	3	8	2	7	9	4	6	5	8	3	2	1	7
5	9	3	2	7	1	4	6	8	1	5	9	7	3	2	6	4	1	9	5	8
6	1	4	3	5	8	9	7	2	4	3	6	1	5	8	2	7	9	3	4	6
						2	5	1	3	9	8	6	7	4						
						3	4	6	2	7	1	8	9	5						
						8	9	7	6	4	5	2	1	3						
9	2	7	4	1	8	6	3	5	9	1	2	4	8	7	3	1	5	2	9	6
3	8	5	6	7	9	1	2	4	7	8	3	5	6	9	2	4	7	1	8	3
1	6	4	5	2	3	7	8	9	5	6	4	3	2	1	6	9	8	5	4	7
2	5	8	9	4	1	3	6	7				8	9	3	1	6	2	4	7	5
4	9	1	3	6	7	8	5	2				2	5	6	7	3	4	9	1	8
6	7	3	8	5	2	4	9	1				1	7	4	5	8	9	3	6	2
7	1	6	2	8	5	9	4	3				7	1	8	4	5	3	6	2	9
8	3	2	1	9	4	5	7	6				6	3	2	9	7	1	8	5	4
5	4	9	7	3	6	2	1	8				9	4	5	8	2	6	7	3	1

27

2	9	4	6	5	7	3	1	8				3	1	9	7	4	8	2	6	5
5	3	6	1	9	8	4	2	7				8	6	5	3	1	2	4	7	9
1	8	7	4	3	2	6	5	9				4	7	2	6	9	5	3	1	8
6	4	3	7	1	5	8	9	2				2	8	3	9	6	4	1	5	7
7	2	5	8	4	9	1	3	6				6	4	1	5	2	7	9	8	3
8	1	9	2	6	3	5	7	4				5	9	7	1	8	3	6	2	4
4	7	1	5	2	6	9	8	3	7	5	6	1	2	4	8	7	9	5	3	6
9	6	8	3	7	1	2	4	5	1	8	9	7	3	6	4	5	1	8	9	2
3	5	2	9	8	4	7	6	1	4	2	3	9	5	8	2	3	6	7	4	1
						3	7	6	9	1	8	2	4	5						
						1	2	4	6	3	5	8	7	9						
						5	9	8	2	4	7	6	1	3						
9	5	6	7	1	8	4	3	2	8	6	1	5	9	7	1	3	4	2	8	6
2	3	1	6	4	9	8	5	7	3	9	2	4	6	1	2	8	7	5	3	9
4	8	7	5	2	3	6	1	9	5	7	4	3	8	2	9	6	5	4	1	7
6	1	5	2	8	4	7	9	3				8	4	3	6	7	1	9	5	2
3	7	4	9	6	5	2	8	1				1	2	5	4	9	8	7	6	3
8	9	2	3	7	1	5	6	4				9	7	6	5	2	3	1	4	8
5	4	9	8	3	7	1	2	6				2	5	8	7	4	6	3	9	1
1	6	3	4	5	2	9	7	8				7	3	4	8	1	9	6	2	5
7	2	8	1	9	6	3	4	5				6	1	9	3	5	2	8	7	4

28

3	9	4	1	6	2	7	8	5				9	2	1	6	3	8	5	7	4
8	1	6	5	7	4	3	9	2				3	8	4	1	7	5	9	2	6
5	7	2	8	3	9	1	6	4				6	5	7	9	2	4	1	3	8
9	3	1	2	8	5	4	7	6				4	1	6	7	9	2	3	8	5
6	2	5	7	4	1	8	3	9				7	3	8	5	4	1	6	9	2
4	8	7	6	9	3	2	5	1				5	9	2	3	8	6	4	1	7
1	6	9	3	2	8	5	4	7	9	1	8	2	6	3	4	1	7	8	5	9
7	5	3	4	1	6	9	2	8	6	4	3	1	7	5	8	6	9	2	4	3
2	4	8	9	5	7	6	1	3	2	7	5	8	4	9	2	5	3	7	6	1
						8	5	9	1	3	7	6	2	4						
						4	6	1	5	2	9	3	8	7						
						7	3	2	8	6	4	9	5	1						
1	5	8	7	2	4	3	9	6	4	5	2	7	1	8	3	2	4	9	6	5
4	6	9	3	8	1	2	7	5	3	8	1	4	9	6	8	5	7	1	2	3
7	3	2	5	6	9	1	8	4	7	9	6	5	3	2	9	1	6	4	8	7
9	4	3	8	7	5	6	2	1				3	5	4	7	8	2	6	1	9
8	1	6	2	4	3	9	5	7				1	6	7	4	9	5	2	3	8
2	7	5	9	1	6	4	3	8				8	2	9	6	3	1	7	5	4
6	2	7	4	3	8	5	1	9				2	8	1	5	4	9	3	7	6
3	9	4	1	5	7	8	6	2				9	7	3	1	6	8	5	4	2
5	8	1	6	9	2	7	4	3				6	4	5	2	7	3	8	9	1

SOLUTIONS

29

1	5	3	2	4	8	6	7	9				2	7	3	5	6	1	8	9	4
2	6	4	9	1	7	5	8	3				6	1	4	3	9	8	2	5	7
9	7	8	5	6	3	2	4	1				8	9	5	4	7	2	6	1	3
4	9	6	8	2	1	3	5	7				3	5	1	7	8	4	9	6	2
5	3	2	4	7	6	1	9	8				9	4	6	2	1	3	7	8	5
7	8	1	3	9	5	4	6	2				7	2	8	6	5	9	3	4	1
6	2	7	1	5	9	8	3	4	2	5	9	1	6	7	9	3	5	4	2	8
8	4	5	7	3	2	9	1	6	4	7	8	5	3	2	8	4	6	1	7	9
3	1	9	6	8	4	7	2	5	1	6	3	4	8	9	1	2	7	5	3	6
						5	8	3	7	9	4	2	1	6						
						2	9	7	6	3	1	8	4	5						
						4	6	1	5	8	2	9	7	3						
4	1	2	6	7	9	3	5	8	9	1	6	7	2	4	1	8	6	5	9	3
6	9	5	8	4	3	1	7	2	3	4	5	6	9	8	7	5	3	4	2	1
8	7	3	1	5	2	6	4	9	8	2	7	3	5	1	4	9	2	7	8	6
3	6	9	2	8	4	7	1	5				2	3	6	8	4	9	1	5	7
1	8	4	7	6	5	9	2	3				5	8	7	3	2	1	9	6	4
5	2	7	9	3	1	8	6	4				1	4	9	6	7	5	8	3	2
7	5	1	4	9	8	2	3	6				8	6	3	9	1	7	2	4	5
2	3	8	5	1	6	4	9	7				9	7	5	2	3	4	6	1	8
9	4	6	3	2	7	5	8	1				4	1	2	5	6	8	3	7	9

30

4	3	8	6	2	7	1	9	5				9	5	1	7	2	3	8	4	6
6	7	9	5	8	1	4	3	2				6	8	2	4	1	9	5	3	7
5	1	2	4	9	3	6	8	7				4	7	3	5	6	8	1	2	9
3	5	6	9	4	2	8	7	1				7	9	5	6	8	4	3	1	2
9	8	4	7	1	5	2	6	3				2	3	8	1	5	7	6	9	4
1	2	7	8	3	6	9	5	4				1	4	6	3	9	2	7	5	8
2	9	5	1	7	8	3	4	6	1	8	9	5	2	7	9	3	6	4	8	1
8	6	1	3	5	4	7	2	9	5	3	6	8	1	4	2	7	5	9	6	3
7	4	3	2	6	9	5	1	8	4	2	7	3	6	9	8	4	1	2	7	5
						4	6	3	7	5	2	1	9	8						
						9	8	1	6	4	3	7	5	2						
						2	5	7	8	9	1	4	3	6						
1	5	2	8	7	9	6	3	4	9	7	5	2	8	1	9	7	5	3	6	4
4	3	7	1	6	5	8	9	2	3	1	4	6	7	5	4	3	2	1	9	8
6	9	8	2	3	4	1	7	5	2	6	8	9	4	3	1	8	6	7	2	5
7	4	6	5	1	8	3	2	9				7	6	9	3	5	8	2	4	1
9	2	1	7	4	3	5	8	6				1	2	8	6	9	4	5	7	3
3	8	5	9	2	6	7	4	1				5	3	4	2	1	7	9	8	6
8	1	4	3	5	2	9	6	7				8	5	6	7	2	1	4	3	9
5	6	3	4	9	7	2	1	8				4	9	7	5	6	3	8	1	2
2	7	9	6	8	1	4	5	3				3	1	2	8	4	9	6	5	7

31

5	7	8	6	4	1	3	2	9				3	4	1	6	9	5	8	2	7
4	6	2	3	8	9	7	5	1				8	6	7	1	2	3	5	9	4
9	3	1	7	5	2	6	8	4				5	9	2	7	4	8	6	1	3
2	9	7	1	3	4	5	6	8				7	1	3	2	6	9	4	5	8
8	1	4	5	6	7	9	3	2				9	2	5	8	7	4	1	3	6
3	5	6	9	2	8	1	4	7				6	8	4	5	3	1	9	7	2
1	2	3	8	7	5	4	9	6	1	3	7	2	5	8	4	1	7	3	6	9
6	4	9	2	1	3	8	7	5	2	4	9	1	3	6	9	8	2	7	4	5
7	8	5	4	9	6	2	1	3	6	8	5	4	7	9	3	5	6	2	8	1
						1	5	7	8	2	6	3	9	4						
						3	6	4	7	9	1	5	8	2						
						9	2	8	3	5	4	7	6	1						
7	4	3	1	9	6	5	8	2	4	6	3	9	1	7	2	4	3	5	8	6
2	8	9	5	4	7	6	3	1	9	7	2	8	4	5	9	7	6	2	1	3
6	1	5	2	8	3	7	4	9	5	1	8	6	2	3	1	5	8	9	4	7
9	7	8	6	5	4	2	1	3				2	9	1	6	8	7	3	5	4
4	6	1	3	2	8	9	5	7				7	5	8	3	2	4	1	6	9
5	3	2	7	1	9	8	6	4				4	3	6	5	9	1	7	2	8
8	2	7	4	3	5	1	9	6				3	8	4	7	1	5	6	9	2
1	5	4	9	6	2	3	7	8				1	6	9	4	3	2	8	7	5
3	9	6	8	7	1	4	2	5				5	7	2	8	6	9	4	3	1

32

3	8	6	9	5	4	2	1	7				1	4	9	2	7	5	3	8	6
4	2	5	6	1	7	8	9	3				5	6	3	4	8	1	7	9	2
7	9	1	8	3	2	6	5	4				2	7	8	6	3	9	1	5	4
8	4	2	7	6	9	1	3	5				9	3	7	8	2	6	4	1	5
5	1	9	4	8	3	7	2	6				4	5	2	9	1	7	6	3	8
6	3	7	5	2	1	4	8	9				8	1	6	5	4	3	2	7	9
1	6	4	3	9	8	5	7	2	3	4	9	6	8	1	7	5	2	9	4	3
9	5	8	2	7	6	3	4	1	8	6	2	7	9	5	3	6	4	8	2	1
2	7	3	1	4	5	9	6	8	1	5	7	3	2	4	1	9	8	5	6	7
						2	3	4	9	8	5	1	6	7						
						7	8	6	4	2	1	5	3	9						
						1	5	9	6	7	3	8	4	2						
6	1	8	2	5	3	4	9	7	5	3	8	2	1	6	9	5	4	3	7	8
9	4	5	7	1	8	6	2	3	7	1	4	9	5	8	1	7	3	6	2	4
2	7	3	4	6	9	8	1	5	2	9	6	4	7	3	8	2	6	1	5	9
8	5	2	3	4	1	7	6	9				6	9	7	3	1	2	8	4	5
7	6	1	5	9	2	3	4	8				5	3	4	6	8	9	2	1	7
3	9	4	6	8	7	2	5	1				1	8	2	7	4	5	9	6	3
5	2	9	8	3	6	1	7	4				7	2	1	4	3	8	5	9	6
1	8	7	9	2	4	5	3	6				8	6	5	2	9	7	4	3	1
4	3	6	1	7	5	9	8	2				3	4	9	5	6	1	7	8	2

SOLUTIONS

33

1	2	7	6	3	9	5	4	8				8	3	7	9	4	1	5	6	2
9	6	5	8	2	4	7	3	1				9	6	2	5	8	7	3	1	4
8	3	4	1	5	7	6	2	9				4	5	1	2	3	6	8	9	7
3	1	9	2	4	5	8	7	6				2	7	9	4	5	8	1	3	6
6	5	2	3	7	8	9	1	4				3	4	6	1	9	2	7	5	8
4	7	8	9	6	1	2	5	3				5	1	8	7	6	3	2	4	9
7	4	6	5	8	3	1	9	2	3	7	4	6	8	5	3	7	9	4	2	1
5	8	1	4	9	2	3	6	7	5	2	8	1	9	4	8	2	5	6	7	3
2	9	3	7	1	6	4	8	5	1	6	9	7	2	3	6	1	4	9	8	5
						8	1	3	7	5	6	9	4	2						
						9	2	6	4	3	1	5	7	8						
						7	5	4	9	8	2	3	1	6						
3	5	1	6	4	8	2	7	9	6	4	5	8	3	1	5	4	7	2	6	9
8	2	6	9	7	3	5	4	1	8	9	3	2	6	7	1	9	3	5	4	8
7	9	4	5	2	1	6	3	8	2	1	7	4	5	9	8	2	6	3	1	7
6	1	8	4	5	7	9	2	3				7	4	5	3	8	1	9	2	6
9	7	5	3	6	2	1	8	4				9	8	3	4	6	2	1	7	5
2	4	3	1	8	9	7	6	5				1	2	6	7	5	9	4	8	3
4	3	9	2	1	6	8	5	7				3	9	2	6	1	8	7	5	4
5	6	7	8	9	4	3	1	2				5	7	8	2	3	4	6	9	1
1	8	2	7	3	5	4	9	6				6	1	4	9	7	5	8	3	2

34

7	1	5	8	6	3	2	9	4				5	3	4	9	2	7	6	1	8
8	6	3	9	4	2	5	7	1				9	8	2	4	1	6	3	5	7
9	2	4	7	5	1	8	6	3				7	6	1	8	3	5	9	4	2
5	9	2	4	3	8	7	1	6				2	5	6	3	9	1	8	7	4
1	4	8	5	7	6	3	2	9				1	4	9	2	7	8	5	3	6
6	3	7	2	1	9	4	5	8				3	7	8	6	5	4	2	9	1
4	5	6	1	8	7	9	3	2	4	7	8	6	1	5	7	8	9	4	2	3
3	8	9	6	2	5	1	4	7	2	6	5	8	9	3	1	4	2	7	6	5
2	7	1	3	9	4	6	8	5	1	9	3	4	2	7	5	6	3	1	8	9
						8	6	9	3	2	7	1	5	4						
						5	1	3	8	4	9	7	6	2						
						2	7	4	5	1	6	9	3	8						
4	6	9	5	8	3	7	2	1	9	5	4	3	8	6	1	4	9	7	5	2
2	3	1	9	7	6	4	5	8	6	3	1	2	7	9	6	3	5	8	4	1
8	7	5	4	2	1	3	9	6	7	8	2	5	4	1	2	7	8	3	6	9
1	9	2	6	4	8	5	3	7				6	1	8	7	9	4	2	3	5
6	4	3	2	5	7	8	1	9				4	5	3	8	6	2	9	1	7
5	8	7	1	3	9	6	4	2				7	9	2	5	1	3	6	8	4
9	5	4	8	6	2	1	7	3				9	6	5	4	8	7	1	2	3
3	2	6	7	1	5	9	8	4				8	3	4	9	2	1	5	7	6
7	1	8	3	9	4	2	6	5				1	2	7	3	5	6	4	9	8

35

4	3	1	7	2	5	9	8	6				5	1	8	9	7	6	2	4	3
5	6	7	8	4	9	3	1	2				7	2	9	3	5	4	8	6	1
8	9	2	1	6	3	4	7	5				6	3	4	2	1	8	9	7	5
2	5	3	6	8	4	7	9	1				9	8	5	6	2	7	3	1	4
7	8	6	9	3	1	5	2	4				3	7	2	1	4	9	5	8	6
9	1	4	5	7	2	6	3	8				4	6	1	5	8	3	7	2	9
1	2	5	3	9	6	8	4	7	3	5	1	2	9	6	7	3	1	4	5	8
3	4	8	2	5	7	1	6	9	4	7	2	8	5	3	4	6	2	1	9	7
6	7	9	4	1	8	2	5	3	9	6	8	1	4	7	8	9	5	6	3	2
						6	7	8	2	9	5	3	1	4						
						3	2	4	1	8	7	9	6	5						
						5	9	1	6	4	3	7	2	8						
2	8	5	9	3	7	4	1	6	7	3	9	5	8	2	1	3	9	4	6	7
9	3	6	4	1	5	7	8	2	5	1	4	6	3	9	4	7	8	2	1	5
4	1	7	2	8	6	9	3	5	8	2	6	4	7	1	2	6	5	9	8	3
1	5	3	7	9	8	2	6	4				2	4	6	7	1	3	5	9	8
8	4	9	1	6	2	5	7	3				7	9	3	5	8	6	1	4	2
7	6	2	5	4	3	1	9	8				8	1	5	9	2	4	7	3	6
6	9	1	3	2	4	8	5	7				3	5	7	6	9	1	8	2	4
5	2	8	6	7	1	3	4	9				9	6	4	8	5	2	3	7	1
3	7	4	8	5	9	6	2	1				1	2	8	3	4	7	6	5	9

36

8	2	6	3	1	5	4	9	7				2	6	7	4	9	1	3	8	5
5	7	4	6	8	9	3	2	1				5	1	4	3	6	8	2	9	7
3	9	1	2	7	4	6	5	8				9	3	8	2	7	5	4	6	1
2	8	5	7	4	3	1	6	9				4	8	5	9	3	7	1	2	6
9	1	3	5	6	2	8	7	4				3	2	1	8	5	6	7	4	9
6	4	7	1	9	8	5	3	2				7	9	6	1	2	4	8	5	3
7	6	2	8	5	1	9	4	3	1	8	5	6	7	2	5	4	3	9	1	8
1	5	9	4	3	7	2	8	6	4	3	7	1	5	9	7	8	2	6	3	4
4	3	8	9	2	6	7	1	5	6	9	2	8	4	3	6	1	9	5	7	2
						5	3	8	7	2	1	4	9	6						
						6	2	9	8	5	4	7	3	1						
						1	7	4	9	6	3	2	8	5						
1	5	3	6	7	4	8	9	2	5	4	6	3	1	7	4	6	2	5	8	9
2	7	8	9	3	5	4	6	1	3	7	9	5	2	8	3	9	7	6	1	4
6	4	9	8	1	2	3	5	7	2	1	8	9	6	4	8	5	1	2	3	7
5	1	4	2	8	6	7	3	9				6	5	1	2	3	4	9	7	8
7	8	6	4	9	3	2	1	5				4	7	3	9	8	6	1	5	2
3	9	2	1	5	7	6	4	8				2	8	9	1	7	5	4	6	3
9	2	5	3	6	8	1	7	4				1	4	6	7	2	8	3	9	5
8	6	1	7	4	9	5	2	3				8	9	2	5	1	3	7	4	6
4	3	7	5	2	1	9	8	6				7	3	5	6	4	9	8	2	1

SOLUTIONS

37

9	5	4	7	6	1	2	8	3				4	1	8	7	5	2	6	9	3
1	7	8	9	3	2	6	4	5				5	7	3	9	6	8	2	4	1
6	3	2	4	5	8	7	1	9				6	9	2	1	3	4	7	8	5
4	9	5	8	7	6	3	2	1				3	8	9	2	7	5	4	1	6
2	1	6	5	9	3	4	7	8				7	6	1	4	8	9	3	5	2
7	8	3	2	1	4	9	5	6				2	4	5	6	1	3	9	7	8
3	2	1	6	4	5	8	9	7	2	3	6	1	5	4	3	9	6	8	2	7
8	6	9	1	2	7	5	3	4	9	7	1	8	2	6	5	4	7	1	3	9
5	4	7	3	8	9	1	6	2	5	8	4	9	3	7	8	2	1	5	6	4
						3	8	1	4	5	7	6	9	2						
						2	4	5	6	1	9	7	8	3						
						6	7	9	8	2	3	4	1	5						
7	3	1	2	4	6	9	5	8	7	4	2	3	6	1	2	8	9	7	4	5
2	5	4	3	8	9	7	1	6	3	9	5	2	4	8	5	1	7	6	3	9
8	9	6	7	1	5	4	2	3	1	6	8	5	7	9	4	6	3	2	1	8
6	7	8	1	3	2	5	4	9				4	3	7	9	2	1	8	5	6
4	2	9	5	6	7	8	3	1				8	1	5	7	4	6	3	9	2
3	1	5	8	9	4	2	6	7				6	9	2	3	5	8	1	7	4
9	6	7	4	5	3	1	8	2				9	8	6	1	3	4	5	2	7
5	8	3	9	2	1	6	7	4				7	2	3	8	9	5	4	6	1
1	4	2	6	7	8	3	9	5				1	5	4	6	7	2	9	8	3

38

5	7	3	2	4	8	9	1	6				8	9	5	3	7	4	2	6	1
8	4	9	1	7	6	2	5	3				2	3	7	1	6	5	8	9	4
6	1	2	5	9	3	7	8	4				4	6	1	8	9	2	5	7	3
2	8	4	6	5	1	3	9	7				3	5	4	7	8	9	6	1	2
1	6	7	9	3	4	8	2	5				6	2	9	4	5	1	7	3	8
3	9	5	8	2	7	6	4	1				1	7	8	2	3	6	4	5	9
7	5	6	4	8	2	1	3	9	2	5	4	7	8	6	9	4	3	1	2	5
4	2	1	3	6	9	5	7	8	1	3	6	9	4	2	5	1	7	3	8	6
9	3	8	7	1	5	4	6	2	7	8	9	5	1	3	6	2	8	9	4	7
						8	5	4	3	7	2	1	6	9						
						9	1	3	4	6	8	2	5	7						
						6	2	7	9	1	5	4	3	8						
5	2	7	9	8	6	3	4	1	8	2	7	6	9	5	2	7	8	1	4	3
9	3	4	5	2	1	7	8	6	5	9	1	3	2	4	1	5	9	8	7	6
1	6	8	7	4	3	2	9	5	6	4	3	8	7	1	4	6	3	5	9	2
8	9	5	3	1	7	4	6	2				9	3	7	8	1	6	2	5	4
2	4	3	8	6	5	9	1	7				1	5	8	9	2	4	6	3	7
6	7	1	4	9	2	5	3	8				4	6	2	5	3	7	9	8	1
4	1	9	2	5	8	6	7	3				5	1	3	7	9	2	4	6	8
7	5	6	1	3	9	8	2	4				7	8	9	6	4	1	3	2	5
3	8	2	6	7	4	1	5	9				2	4	6	3	8	5	7	1	9

39

7	8	2	1	6	4	5	9	3				8	7	4	2	3	5	6	1	9
3	9	1	7	2	5	8	6	4				5	1	3	6	9	7	8	2	4
5	4	6	9	8	3	2	1	7				2	6	9	1	4	8	5	7	3
9	1	4	5	7	6	3	2	8				4	5	8	9	7	2	1	3	6
8	5	3	4	1	2	9	7	6				6	2	7	3	8	1	9	4	5
6	2	7	3	9	8	4	5	1				3	9	1	4	5	6	7	8	2
2	3	8	6	5	1	7	4	9	2	5	3	1	8	6	5	2	4	3	9	7
4	6	9	2	3	7	1	8	5	6	9	4	7	3	2	8	6	9	4	5	1
1	7	5	8	4	9	6	3	2	8	1	7	9	4	5	7	1	3	2	6	8
						2	7	4	3	6	9	8	5	1						
						5	6	3	4	8	1	2	7	9						
						8	9	1	7	2	5	3	6	4						
9	5	7	8	3	2	4	1	6	9	7	8	5	2	3	1	4	7	6	8	9
2	4	8	9	6	1	3	5	7	1	4	2	6	9	8	5	3	2	1	4	7
1	3	6	5	4	7	9	2	8	5	3	6	4	1	7	8	6	9	5	3	2
7	1	4	2	5	3	8	6	9				9	7	6	3	8	5	4	2	1
8	2	5	7	9	6	1	4	3				2	4	5	7	1	6	8	9	3
3	6	9	1	8	4	5	7	2				8	3	1	9	2	4	7	6	5
6	8	2	3	1	5	7	9	4				7	8	2	6	9	1	3	5	4
4	9	1	6	7	8	2	3	5				3	5	9	4	7	8	2	1	6
5	7	3	4	2	9	6	8	1				1	6	4	2	5	3	9	7	8

40

6	4	8	7	2	5	9	1	3				4	9	7	2	6	8	5	1	3
9	5	1	4	8	3	2	7	6				5	3	6	1	9	4	8	7	2
7	3	2	1	9	6	5	4	8				2	1	8	3	5	7	4	6	9
4	1	6	8	7	9	3	5	2				6	4	2	8	3	9	7	5	1
5	2	3	6	4	1	8	9	7				9	5	1	7	4	2	3	8	6
8	7	9	3	5	2	4	6	1				7	8	3	5	1	6	9	2	4
1	9	5	2	3	7	6	8	4	9	7	1	3	2	5	4	8	1	6	9	7
2	8	7	9	6	4	1	3	5	4	6	2	8	7	9	6	2	3	1	4	5
3	6	4	5	1	8	7	2	9	8	5	3	1	6	4	9	7	5	2	3	8
						4	5	2	3	9	7	6	8	1						
						8	1	7	6	2	5	4	9	3						
						3	9	6	1	8	4	7	5	2						
9	1	3	5	7	6	2	4	8	7	3	9	5	1	6	9	4	2	3	8	7
2	4	6	3	9	8	5	7	1	2	4	6	9	3	8	5	1	7	4	6	2
8	7	5	1	2	4	9	6	3	5	1	8	2	4	7	6	8	3	5	9	1
4	6	2	7	8	3	1	9	5				7	9	4	3	5	6	2	1	8
3	9	8	4	1	5	6	2	7				8	5	3	2	7	1	6	4	9
7	5	1	9	6	2	8	3	4				1	6	2	4	9	8	7	5	3
6	3	4	2	5	1	7	8	9				3	2	1	8	6	4	9	7	5
1	2	9	8	3	7	4	5	6				6	8	5	7	2	9	1	3	4
5	8	7	6	4	9	3	1	2				4	7	9	1	3	5	8	2	6

SOLUTIONS

41

4	1	8	5	2	7	6	3	9				9	8	1	5	7	2	4	3	6
5	9	7	8	3	6	2	1	4				3	6	5	1	9	4	8	2	7
3	2	6	9	4	1	8	5	7				7	4	2	8	6	3	5	1	9
6	4	9	2	7	5	1	8	3				4	1	6	9	3	5	2	7	8
7	3	2	1	8	4	9	6	5				5	2	3	7	8	1	6	9	4
8	5	1	6	9	3	7	4	2				8	7	9	2	4	6	1	5	3
9	7	4	3	1	8	5	2	6	4	3	7	1	9	8	4	5	7	3	6	2
1	6	3	7	5	2	4	9	8	5	6	1	2	3	7	6	1	8	9	4	5
2	8	5	4	6	9	3	7	1	8	2	9	6	5	4	3	2	9	7	8	1
						7	8	3	2	4	6	5	1	9						
						9	4	2	1	7	5	8	6	3						
						6	1	5	3	9	8	7	4	2						
9	7	5	8	2	3	1	6	4	9	8	2	3	7	5	8	6	2	1	9	4
6	2	3	4	7	1	8	5	9	7	1	3	4	2	6	3	1	9	5	8	7
4	1	8	5	9	6	2	3	7	6	5	4	9	8	1	4	5	7	6	3	2
8	4	2	3	5	7	6	9	1				2	5	9	1	8	6	7	4	3
3	5	6	9	1	4	7	8	2				8	1	4	7	3	5	2	6	9
1	9	7	2	6	8	3	4	5				7	6	3	2	9	4	8	5	1
7	3	9	1	8	5	4	2	6				5	4	7	6	2	3	9	1	8
2	6	4	7	3	9	5	1	8				6	3	8	9	7	1	4	2	5
5	8	1	6	4	2	9	7	3				1	9	2	5	4	8	3	7	6

42

4	7	6	1	9	2	5	3	8				6	3	2	1	8	4	9	7	5
2	3	9	8	6	5	1	7	4				8	9	4	7	2	5	6	3	1
5	1	8	4	3	7	6	9	2				7	5	1	9	6	3	8	4	2
3	9	2	5	8	1	7	4	6				2	8	7	3	9	6	1	5	4
7	4	5	9	2	6	8	1	3				9	1	6	5	4	2	7	8	3
6	8	1	3	7	4	2	5	9				5	4	3	8	1	7	2	6	9
8	6	3	7	1	9	4	2	5	7	8	1	3	6	9	4	7	1	5	2	8
1	2	4	6	5	3	9	8	7	6	3	4	1	2	5	6	3	8	4	9	7
9	5	7	2	4	8	3	6	1	9	5	2	4	7	8	2	5	9	3	1	6
						1	9	2	3	7	8	5	4	6						
						6	4	8	2	9	5	7	1	3						
						7	5	3	1	4	6	9	8	2						
2	7	3	8	6	9	5	1	4	8	2	9	6	3	7	9	2	5	1	4	8
9	5	4	2	3	1	8	7	6	5	1	3	2	9	4	1	3	8	6	7	5
8	6	1	7	5	4	2	3	9	4	6	7	8	5	1	6	4	7	9	2	3
4	9	8	1	2	3	7	6	5				1	8	3	2	9	4	7	5	6
1	2	5	4	7	6	9	8	3				5	6	9	7	1	3	4	8	2
7	3	6	5	9	8	1	4	2				7	4	2	8	5	6	3	1	9
6	4	7	9	8	5	3	2	1				4	1	8	5	6	9	2	3	7
5	1	2	3	4	7	6	9	8				9	2	5	3	7	1	8	6	4
3	8	9	6	1	2	4	5	7				3	7	6	4	8	2	5	9	1

43

6	4	1	3	8	7	2	5	9				8	6	4	7	1	9	2	3	5
3	9	2	5	6	1	8	4	7				1	9	2	3	6	5	7	4	8
7	5	8	2	4	9	6	3	1				7	3	5	4	2	8	6	9	1
5	6	4	1	3	8	7	9	2				5	1	6	2	7	3	4	8	9
2	8	3	7	9	5	4	1	6				4	7	9	5	8	1	3	6	2
9	1	7	4	2	6	3	8	5				2	8	3	6	9	4	5	1	7
8	2	5	9	7	4	1	6	3	8	4	5	9	2	7	1	3	6	8	5	4
1	7	6	8	5	3	9	2	4	3	1	7	6	5	8	9	4	2	1	7	3
4	3	9	6	1	2	5	7	8	9	2	6	3	4	1	8	5	7	9	2	6
						3	5	2	7	6	1	4	8	9						
						4	1	7	5	9	8	2	3	6						
						6	8	9	2	3	4	1	7	5						
3	5	1	4	2	8	7	9	6	4	5	3	8	1	2	7	4	5	3	6	9
7	4	8	1	6	9	2	3	5	1	8	9	7	6	4	3	9	8	5	1	2
2	6	9	5	7	3	8	4	1	6	7	2	5	9	3	6	2	1	7	4	8
9	2	5	3	4	1	6	7	8				2	8	6	5	3	4	1	9	7
4	8	3	6	5	7	9	1	2				4	5	7	1	6	9	8	2	3
6	1	7	9	8	2	4	5	3				9	3	1	2	8	7	6	5	4
5	3	4	2	9	6	1	8	7				1	2	9	8	7	6	4	3	5
8	9	6	7	1	5	3	2	4				3	7	5	4	1	2	9	8	6
1	7	2	8	3	4	5	6	9				6	4	8	9	5	3	2	7	1

44

3	1	8	4	6	9	5	2	7				7	5	1	2	9	6	4	3	8
5	6	4	2	7	8	9	3	1				4	9	2	1	8	3	7	6	5
2	7	9	3	5	1	4	6	8				8	6	3	4	5	7	2	1	9
4	2	3	5	9	7	8	1	6				5	7	9	3	6	2	1	8	4
1	8	5	6	4	2	7	9	3				6	1	8	5	7	4	9	2	3
6	9	7	1	8	3	2	5	4				2	3	4	8	1	9	6	5	7
8	5	6	9	3	4	1	7	2	5	4	3	9	8	6	7	2	5	3	4	1
7	3	1	8	2	5	6	4	9	7	1	8	3	2	5	9	4	1	8	7	6
9	4	2	7	1	6	3	8	5	2	9	6	1	4	7	6	3	8	5	9	2
						9	3	1	8	5	4	7	6	2						
						4	5	7	6	3	2	8	1	9						
						2	6	8	9	7	1	5	3	4						
7	6	4	8	9	1	5	2	3	1	6	7	4	9	8	3	5	1	7	6	2
5	3	1	7	6	2	8	9	4	3	2	5	6	7	1	8	2	4	9	5	3
2	9	8	3	5	4	7	1	6	4	8	9	2	5	3	7	9	6	1	8	4
9	2	5	1	3	6	4	8	7				9	4	6	2	1	5	3	7	8
1	4	7	2	8	5	6	3	9				1	2	7	6	8	3	5	4	9
6	8	3	4	7	9	1	5	2				3	8	5	4	7	9	2	1	6
3	7	6	5	2	8	9	4	1				7	6	4	1	3	2	8	9	5
4	5	2	9	1	7	3	6	8				8	3	9	5	6	7	4	2	1
8	1	9	6	4	3	2	7	5				5	1	2	9	4	8	6	3	7

SOLUTIONS

45

7	8	6	4	2	9	1	5	3				9	1	5	2	7	6	8	3	4
1	9	4	3	5	7	2	8	6				3	7	2	4	1	8	9	6	5
5	2	3	8	6	1	4	7	9				8	6	4	3	9	5	1	7	2
4	1	7	6	3	8	9	2	5				4	8	7	5	6	1	2	9	3
2	6	9	7	1	5	8	3	4				1	2	9	8	4	3	7	5	6
8	3	5	2	9	4	6	1	7				5	3	6	9	2	7	4	8	1
6	5	1	9	8	3	7	4	2	3	8	5	6	9	1	7	5	2	3	4	8
9	4	8	5	7	2	3	6	1	2	4	9	7	5	8	1	3	4	6	2	9
3	7	2	1	4	6	5	9	8	7	6	1	2	4	3	6	8	9	5	1	7
						1	8	4	6	9	2	5	3	7						
						6	7	5	8	1	3	4	2	9						
						9	2	3	4	5	7	8	1	6						
1	6	7	4	5	8	2	3	9	5	7	6	1	8	4	3	9	6	5	2	7
3	4	2	9	1	6	8	5	7	1	3	4	9	6	2	1	7	5	3	8	4
9	8	5	2	7	3	4	1	6	9	2	8	3	7	5	2	8	4	6	1	9
4	1	3	7	8	2	9	6	5				5	9	1	4	6	3	2	7	8
6	5	8	1	3	9	7	4	2				4	3	6	7	2	8	9	5	1
2	7	9	5	6	4	3	8	1				8	2	7	9	5	1	4	6	3
8	2	6	3	9	1	5	7	4				6	5	3	8	4	7	1	9	2
7	3	4	6	2	5	1	9	8				7	4	9	6	1	2	8	3	5
5	9	1	8	4	7	6	2	3				2	1	8	5	3	9	7	4	6

46

6	2	7	8	4	3	9	5	1				5	1	8	7	9	2	4	3	6
5	3	8	1	9	2	6	4	7				3	9	7	4	6	8	2	1	5
4	9	1	7	5	6	3	2	8				4	2	6	5	3	1	8	9	7
8	6	3	4	7	1	5	9	2				9	7	1	3	5	4	6	2	8
9	7	4	2	3	5	8	1	6				8	3	4	9	2	6	7	5	1
2	1	5	9	6	8	4	7	3				2	6	5	1	8	7	9	4	3
3	8	2	5	1	4	7	6	9	2	8	5	1	4	3	8	7	9	5	6	2
1	4	9	6	8	7	2	3	5	1	7	4	6	8	9	2	1	5	3	7	4
7	5	6	3	2	9	1	8	4	3	9	6	7	5	2	6	4	3	1	8	9
						8	7	3	4	5	1	2	9	6						
						4	1	6	9	2	7	8	3	5						
						9	5	2	6	3	8	4	1	7						
5	4	1	8	6	9	3	2	7	8	1	9	5	6	4	7	8	9	1	3	2
3	8	2	7	4	5	6	9	1	5	4	2	3	7	8	1	2	5	9	6	4
9	7	6	2	1	3	5	4	8	7	6	3	9	2	1	3	4	6	5	7	8
1	5	4	9	7	6	2	8	3				2	3	9	6	1	7	4	8	5
2	3	7	5	8	1	4	6	9				1	8	7	4	5	2	3	9	6
6	9	8	4	3	2	7	1	5				6	4	5	8	9	3	2	1	7
8	2	9	6	5	7	1	3	4				8	5	2	9	7	1	6	4	3
7	6	3	1	9	4	8	5	2				7	1	6	2	3	4	8	5	9
4	1	5	3	2	8	9	7	6				4	9	3	5	6	8	7	2	1

47

7	8	6	3	1	9	5	4	2				5	4	8	6	9	2	1	3	7
9	1	3	5	2	4	7	6	8				3	6	2	7	1	5	8	4	9
2	4	5	6	7	8	3	1	9				7	9	1	3	4	8	6	5	2
5	9	2	7	8	1	6	3	4				6	8	4	1	3	9	2	7	5
1	6	4	9	3	5	8	2	7				1	5	3	4	2	7	9	8	6
8	3	7	2	4	6	9	5	1				2	7	9	5	8	6	3	1	4
6	7	9	1	5	2	4	8	3	1	6	5	9	2	7	8	5	1	4	6	3
3	2	8	4	6	7	1	9	5	2	4	7	8	3	6	9	7	4	5	2	1
4	5	1	8	9	3	2	7	6	8	9	3	4	1	5	2	6	3	7	9	8
						3	2	9	4	1	6	5	7	8						
						8	4	1	7	5	2	3	6	9						
						5	6	7	9	3	8	1	4	2						
4	8	6	5	3	9	7	1	2	5	8	4	6	9	3	7	8	1	5	4	2
7	1	5	4	2	6	9	3	8	6	2	1	7	5	4	6	3	2	8	1	9
3	9	2	8	7	1	6	5	4	3	7	9	2	8	1	9	4	5	6	7	3
9	4	7	1	8	3	5	2	6				1	2	9	4	5	7	3	8	6
2	5	3	9	6	4	8	7	1				8	4	6	3	1	9	7	2	5
1	6	8	7	5	2	3	4	9				3	7	5	8	2	6	4	9	1
6	7	9	2	4	5	1	8	3				9	1	7	5	6	4	2	3	8
8	2	1	3	9	7	4	6	5				4	6	8	2	9	3	1	5	7
5	3	4	6	1	8	2	9	7				5	3	2	1	7	8	9	6	4

48

1	8	9	5	2	4	3	6	7				1	6	9	5	4	7	3	2	8
6	4	3	7	8	9	1	2	5				8	2	4	6	9	3	5	7	1
7	2	5	3	6	1	9	8	4				3	7	5	8	2	1	9	6	4
8	7	2	1	3	6	5	4	9				2	5	6	1	3	9	8	4	7
4	5	6	8	9	7	2	1	3				4	9	1	2	7	8	6	3	5
9	3	1	4	5	2	6	7	8				7	3	8	4	6	5	1	9	2
5	9	7	6	1	8	4	3	2	5	8	6	9	1	7	3	5	4	2	8	6
3	1	4	2	7	5	8	9	6	1	3	7	5	4	2	9	8	6	7	1	3
2	6	8	9	4	3	7	5	1	2	4	9	6	8	3	7	1	2	4	5	9
						6	4	5	7	9	3	1	2	8						
						2	7	9	4	1	8	3	6	5						
						1	8	3	6	5	2	4	7	9						
8	2	4	3	9	1	5	6	7	9	2	4	8	3	1	9	6	5	4	7	2
7	9	6	8	5	2	3	1	4	8	7	5	2	9	6	7	3	4	1	8	5
5	3	1	6	7	4	9	2	8	3	6	1	7	5	4	1	8	2	3	6	9
9	8	5	7	4	6	1	3	2				3	7	9	5	4	1	6	2	8
3	6	7	2	1	5	4	8	9				1	8	5	3	2	6	7	9	4
4	1	2	9	3	8	7	5	6				4	6	2	8	7	9	5	1	3
2	5	3	4	6	7	8	9	1				9	1	8	6	5	3	2	4	7
1	4	8	5	2	9	6	7	3				5	4	7	2	1	8	9	3	6
6	7	9	1	8	3	2	4	5				6	2	3	4	9	7	8	5	1

SOLUTIONS

49

9	6	7	2	5	8	3	4	1				5	8	4	9	7	6	2	1	3
1	4	2	7	6	3	9	8	5				6	2	3	5	1	8	4	7	9
8	3	5	1	9	4	7	2	6				7	1	9	4	3	2	6	5	8
3	9	1	6	8	5	2	7	4				3	7	1	8	5	4	9	2	6
2	7	6	9	4	1	5	3	8				9	4	5	2	6	7	3	8	1
4	5	8	3	2	7	1	6	9				2	6	8	1	9	3	5	4	7
7	2	9	4	1	6	8	5	3	6	1	2	4	9	7	3	2	1	8	6	5
5	1	4	8	3	2	6	9	7	8	5	4	1	3	2	6	8	5	7	9	4
6	8	3	5	7	9	4	1	2	9	3	7	8	5	6	7	4	9	1	3	2
						2	4	9	7	8	1	5	6	3						
						1	7	8	3	6	5	2	4	9						
						5	3	6	2	4	9	7	8	1						
8	7	2	5	1	9	3	6	4	1	2	8	9	7	5	8	4	1	3	2	6
3	5	9	6	4	2	7	8	1	5	9	6	3	2	4	9	7	6	8	5	1
1	4	6	8	7	3	9	2	5	4	7	3	6	1	8	5	2	3	7	9	4
2	3	1	4	6	5	8	9	7				4	6	7	2	9	8	1	3	5
9	8	7	3	2	1	5	4	6				1	3	2	7	6	5	9	4	8
5	6	4	7	9	8	2	1	3				5	8	9	3	1	4	2	6	7
6	2	3	9	5	4	1	7	8				8	5	1	4	3	9	6	7	2
7	1	8	2	3	6	4	5	9				2	9	6	1	5	7	4	8	3
4	9	5	1	8	7	6	3	2				7	4	3	6	8	2	5	1	9

50

5	6	8	4	7	2	3	9	1				3	7	5	1	4	6	2	9	8
9	1	4	8	6	3	5	7	2				8	4	2	9	5	7	1	6	3
2	7	3	9	5	1	4	8	6				1	9	6	3	8	2	4	5	7
8	9	6	2	3	7	1	4	5				2	5	9	7	1	8	6	3	4
7	5	1	6	9	4	8	2	3				4	6	1	5	2	3	7	8	9
4	3	2	5	1	8	9	6	7				7	3	8	6	9	4	5	2	1
6	2	9	3	8	5	7	1	4	2	5	9	6	8	3	2	7	1	9	4	5
1	8	5	7	4	6	2	3	9	8	6	7	5	1	4	8	6	9	3	7	2
3	4	7	1	2	9	6	5	8	1	3	4	9	2	7	4	3	5	8	1	6
						1	4	7	9	8	6	2	3	5						
						5	9	6	4	2	3	8	7	1						
						3	8	2	7	1	5	4	9	6						
9	1	2	6	5	8	4	7	3	5	9	8	1	6	2	7	5	3	8	4	9
3	6	7	9	4	1	8	2	5	6	7	1	3	4	9	8	1	6	7	2	5
5	4	8	3	2	7	9	6	1	3	4	2	7	5	8	2	4	9	1	3	6
7	9	3	2	6	5	1	4	8				4	9	5	1	3	2	6	7	8
6	8	1	4	7	9	3	5	2				8	1	7	4	6	5	2	9	3
4	2	5	1	8	3	6	9	7				6	2	3	9	8	7	4	5	1
8	3	6	7	9	2	5	1	4				5	3	4	6	7	8	9	1	2
1	7	9	5	3	4	2	8	6				9	7	6	5	2	1	3	8	4
2	5	4	8	1	6	7	3	9				2	8	1	3	9	4	5	6	7

51

6	8	9	3	7	5	2	4	1				6	8	9	2	1	5	4	7	3
4	1	5	2	8	6	7	9	3				3	4	5	7	9	6	1	2	8
2	7	3	1	9	4	5	6	8				2	7	1	4	3	8	6	9	5
7	6	8	9	2	3	1	5	4				7	9	8	5	6	4	3	1	2
1	5	2	6	4	7	3	8	9				4	5	6	1	2	3	9	8	7
9	3	4	5	1	8	6	7	2				1	3	2	8	7	9	5	6	4
8	9	1	7	6	2	4	3	5	2	8	1	9	6	7	3	5	2	8	4	1
5	2	6	4	3	9	8	1	7	9	6	3	5	2	4	6	8	1	7	3	9
3	4	7	8	5	1	9	2	6	5	7	4	8	1	3	9	4	7	2	5	6
						3	7	4	6	2	8	1	9	5						
						1	8	9	7	4	5	6	3	2						
						6	5	2	3	1	9	7	4	8						
8	2	9	4	1	5	7	6	3	1	5	2	4	8	9	5	7	6	3	2	1
7	6	5	8	9	3	2	4	1	8	9	7	3	5	6	1	2	8	4	7	9
1	4	3	6	7	2	5	9	8	4	3	6	2	7	1	3	4	9	5	8	6
4	7	8	3	5	9	1	2	6				8	4	7	2	1	5	6	9	3
3	5	1	2	4	6	9	8	7				1	6	5	9	8	3	7	4	2
2	9	6	7	8	1	4	3	5				9	3	2	7	6	4	8	1	5
5	3	2	9	6	7	8	1	4				7	2	3	8	5	1	9	6	4
6	1	4	5	2	8	3	7	9				6	9	8	4	3	2	1	5	7
9	8	7	1	3	4	6	5	2				5	1	4	6	9	7	2	3	8

52

1	6	5	3	2	9	7	8	4				5	7	2	3	6	4	1	9	8
3	4	8	7	1	5	9	2	6				9	8	1	2	7	5	6	4	3
2	7	9	8	4	6	3	1	5				3	6	4	8	1	9	7	5	2
7	8	3	5	6	1	4	9	2				1	3	6	9	4	8	2	7	5
6	9	1	4	7	2	8	5	3				7	5	8	1	2	3	9	6	4
4	5	2	9	8	3	6	7	1				2	4	9	6	5	7	3	8	1
5	2	7	6	9	4	1	3	8	9	4	7	6	2	5	4	9	1	8	3	7
8	1	6	2	3	7	5	4	9	3	6	2	8	1	7	5	3	6	4	2	9
9	3	4	1	5	8	2	6	7	5	1	8	4	9	3	7	8	2	5	1	6
						9	5	1	8	3	4	7	6	2						
						6	8	4	2	7	5	9	3	1						
						3	7	2	1	9	6	5	4	8						
3	4	8	6	1	2	7	9	5	4	2	1	3	8	6	4	2	5	7	9	1
5	6	1	9	7	4	8	2	3	6	5	9	1	7	4	8	3	9	2	5	6
2	9	7	8	5	3	4	1	6	7	8	3	2	5	9	6	1	7	3	8	4
7	5	2	4	6	9	1	3	8				8	3	2	7	6	4	5	1	9
4	1	3	7	2	8	6	5	9				7	6	1	5	9	2	8	4	3
6	8	9	5	3	1	2	7	4				4	9	5	3	8	1	6	7	2
1	2	5	3	8	6	9	4	7				6	4	3	9	7	8	1	2	5
9	7	6	1	4	5	3	8	2				9	2	8	1	5	3	4	6	7
8	3	4	2	9	7	5	6	1				5	1	7	2	4	6	9	3	8

SOLUTIONS

53

2	8	4	6	7	9	1	5	3				2	5	7	3	4	1	8	9	6
6	1	7	3	5	4	9	8	2				4	9	6	8	5	2	1	7	3
5	9	3	8	1	2	4	6	7				8	3	1	6	9	7	4	5	2
7	5	2	4	8	1	3	9	6				3	1	4	9	8	6	5	2	7
1	6	8	7	9	3	2	4	5				9	8	2	1	7	5	3	6	4
4	3	9	5	2	6	7	1	8				6	7	5	2	3	4	9	1	8
9	2	6	1	3	5	8	7	4	9	2	1	5	6	3	7	1	8	2	4	9
8	4	1	2	6	7	5	3	9	6	7	4	1	2	8	4	6	9	7	3	5
3	7	5	9	4	8	6	2	1	8	5	3	7	4	9	5	2	3	6	8	1
						2	4	6	3	1	7	9	8	5						
						7	1	8	5	6	9	2	3	4						
						3	9	5	4	8	2	6	1	7						
5	9	7	8	4	2	1	6	3	7	4	5	8	9	2	3	7	4	1	6	5
6	4	8	3	7	1	9	5	2	1	3	8	4	7	6	1	5	2	9	8	3
2	3	1	6	9	5	4	8	7	2	9	6	3	5	1	6	8	9	2	7	4
4	5	2	7	6	9	3	1	8				1	8	7	5	4	3	6	2	9
1	8	6	5	3	4	2	7	9				9	4	5	8	2	6	3	1	7
9	7	3	1	2	8	6	4	5				6	2	3	9	1	7	5	4	8
8	2	5	9	1	6	7	3	4				2	6	9	4	3	8	7	5	1
7	1	9	4	8	3	5	2	6				5	3	4	7	6	1	8	9	2
3	6	4	2	5	7	8	9	1				7	1	8	2	9	5	4	3	6

54

5	7	1	6	2	8	4	3	9				3	8	4	5	9	7	2	6	1
9	2	3	5	7	4	6	1	8				1	7	2	4	8	6	5	3	9
8	4	6	1	9	3	2	7	5				9	6	5	2	3	1	7	8	4
1	5	7	3	4	6	9	8	2				5	1	7	9	2	8	6	4	3
2	8	4	9	1	5	7	6	3				2	3	6	1	5	4	8	9	7
6	3	9	2	8	7	1	5	4				4	9	8	7	6	3	1	2	5
7	6	8	4	3	9	5	2	1	3	8	7	6	4	9	8	1	5	3	7	2
4	1	5	8	6	2	3	9	7	6	4	5	8	2	1	3	7	9	4	5	6
3	9	2	7	5	1	8	4	6	9	2	1	7	5	3	6	4	2	9	1	8
						7	8	9	4	1	6	5	3	2						
						6	3	5	7	9	2	1	8	4						
						2	1	4	5	3	8	9	7	6						
7	5	2	4	3	9	1	6	8	2	7	3	4	9	5	6	2	3	1	8	7
9	6	8	1	5	2	4	7	3	1	5	9	2	6	8	5	1	7	4	3	9
1	4	3	6	8	7	9	5	2	8	6	4	3	1	7	8	4	9	5	6	2
2	9	6	7	4	3	8	1	5				9	4	1	2	8	6	3	7	5
8	3	1	2	6	5	7	9	4				6	7	3	1	9	5	8	2	4
4	7	5	9	1	8	2	3	6				8	5	2	3	7	4	6	9	1
5	8	7	3	2	1	6	4	9				1	2	9	4	3	8	7	5	6
6	2	9	5	7	4	3	8	1				7	8	6	9	5	1	2	4	3
3	1	4	8	9	6	5	2	7				5	3	4	7	6	2	9	1	8

55

9	6	2	3	4	7	8	5	1				4	2	6	7	5	3	9	8	1
8	5	1	9	6	2	3	4	7				9	1	5	2	4	8	6	7	3
7	4	3	1	8	5	6	2	9				7	8	3	9	1	6	5	4	2
5	9	7	6	3	1	2	8	4				6	3	9	8	7	1	2	5	4
4	3	8	7	2	9	1	6	5				5	7	1	6	2	4	3	9	8
1	2	6	8	5	4	7	9	3				2	4	8	3	9	5	7	1	6
2	7	5	4	1	8	9	3	6	4	7	8	1	5	2	4	3	7	8	6	9
6	1	4	2	9	3	5	7	8	9	1	2	3	6	4	5	8	9	1	2	7
3	8	9	5	7	6	4	1	2	3	6	5	8	9	7	1	6	2	4	3	5
						8	2	5	6	4	7	9	1	3						
						6	9	1	8	2	3	7	4	5						
						7	4	3	5	9	1	6	2	8						
9	3	5	4	1	6	2	8	7	1	5	9	4	3	6	5	2	9	7	1	8
4	7	1	2	8	5	3	6	9	2	8	4	5	7	1	8	4	3	9	6	2
8	2	6	3	9	7	1	5	4	7	3	6	2	8	9	1	6	7	5	4	3
2	6	7	9	3	8	4	1	5				3	1	8	2	7	5	4	9	6
3	4	9	7	5	1	6	2	8				9	4	2	6	3	1	8	5	7
5	1	8	6	4	2	7	9	3				6	5	7	9	8	4	2	3	1
6	8	4	5	2	3	9	7	1				1	2	5	3	9	8	6	7	4
7	5	3	1	6	9	8	4	2				7	6	3	4	5	2	1	8	9
1	9	2	8	7	4	5	3	6				8	9	4	7	1	6	3	2	5

56

5	6	9	4	2	3	7	1	8				3	9	1	2	7	4	8	5	6
8	4	1	7	9	5	2	6	3				6	8	4	5	9	3	2	7	1
7	3	2	1	8	6	5	9	4				7	2	5	8	1	6	3	9	4
2	5	6	9	3	1	4	8	7				8	6	9	1	5	2	4	3	7
3	7	4	2	6	8	9	5	1				1	3	7	4	6	8	5	2	9
1	9	8	5	4	7	3	2	6				4	5	2	9	3	7	6	1	8
4	1	7	8	5	2	6	3	9	1	2	7	5	4	8	7	2	1	9	6	3
9	2	3	6	1	4	8	7	5	9	6	4	2	1	3	6	4	9	7	8	5
6	8	5	3	7	9	1	4	2	8	3	5	9	7	6	3	8	5	1	4	2
						9	2	4	6	7	1	8	3	5						
						7	6	3	2	5	8	1	9	4						
						5	8	1	4	9	3	7	6	2						
6	7	5	3	2	1	4	9	8	3	1	2	6	5	7	2	3	4	1	8	9
4	9	2	5	6	8	3	1	7	5	8	6	4	2	9	6	1	8	5	3	7
8	3	1	4	7	9	2	5	6	7	4	9	3	8	1	9	7	5	2	6	4
2	8	4	9	5	3	6	7	1				5	6	4	1	9	7	3	2	8
9	1	6	7	4	2	8	3	5				1	3	8	4	2	6	7	9	5
3	5	7	1	8	6	9	2	4				9	7	2	5	8	3	6	4	1
1	4	9	8	3	5	7	6	2				8	4	3	7	5	2	9	1	6
5	2	8	6	9	7	1	4	3				2	1	5	8	6	9	4	7	3
7	6	3	2	1	4	5	8	9				7	9	6	3	4	1	8	5	2

SOLUTIONS

57

5	3	6	7	2	8	4	9	1				1	9	5	3	6	2	7	4	8
9	2	1	6	3	4	7	5	8				3	8	2	5	7	4	6	9	1
7	4	8	9	5	1	3	6	2				6	4	7	8	1	9	2	5	3
6	1	7	3	9	2	8	4	5				4	1	3	7	9	6	8	2	5
2	9	5	4	8	7	6	1	3				2	5	9	1	8	3	4	7	6
4	8	3	5	1	6	2	7	9				7	6	8	4	2	5	1	3	9
1	6	9	8	7	3	5	2	4	1	8	3	9	7	6	2	3	8	5	1	4
8	5	4	2	6	9	1	3	7	9	6	5	8	2	4	9	5	1	3	6	7
3	7	2	1	4	5	9	8	6	7	2	4	5	3	1	6	4	7	9	8	2
						6	7	3	2	1	9	4	8	5						
						2	9	8	4	5	6	7	1	3						
						4	1	5	3	7	8	6	9	2						
4	3	1	5	2	7	8	6	9	5	3	2	1	4	7	2	8	5	6	9	3
2	9	7	6	4	8	3	5	1	8	4	7	2	6	9	1	3	7	4	8	5
6	5	8	1	9	3	7	4	2	6	9	1	3	5	8	6	4	9	1	7	2
3	6	5	7	8	1	9	2	4				7	1	3	8	6	4	5	2	9
8	7	4	2	6	9	1	3	5				6	2	5	9	7	1	3	4	8
1	2	9	3	5	4	6	8	7				8	9	4	5	2	3	7	6	1
9	4	3	8	7	5	2	1	6				5	7	1	4	9	8	2	3	6
5	8	2	9	1	6	4	7	3				9	3	6	7	5	2	8	1	4
7	1	6	4	3	2	5	9	8				4	8	2	3	1	6	9	5	7

58

5	2	4	6	9	3	7	1	8				5	9	3	1	6	2	7	8	4
3	8	6	4	7	1	5	2	9				8	6	1	4	9	7	3	2	5
1	9	7	8	2	5	3	6	4				4	2	7	3	5	8	6	1	9
9	5	1	7	4	2	8	3	6				9	7	5	6	3	1	8	4	2
6	7	3	9	5	8	1	4	2				3	4	8	5	2	9	1	6	7
8	4	2	1	3	6	9	5	7				2	1	6	7	8	4	5	9	3
2	1	9	3	6	7	4	8	5	7	1	2	6	3	9	8	4	5	2	7	1
4	6	8	5	1	9	2	7	3	6	8	9	1	5	4	2	7	6	9	3	8
7	3	5	2	8	4	6	9	1	5	4	3	7	8	2	9	1	3	4	5	6
						9	6	7	1	5	8	2	4	3						
						1	5	2	3	6	4	8	9	7						
						3	4	8	2	9	7	5	1	6						
1	5	7	4	2	9	8	3	6	4	2	5	9	7	1	2	6	8	5	3	4
4	3	8	5	1	6	7	2	9	8	3	1	4	6	5	9	1	3	8	2	7
9	2	6	3	7	8	5	1	4	9	7	6	3	2	8	5	7	4	9	1	6
6	9	1	7	8	3	2	4	5				6	3	2	4	5	7	1	9	8
7	4	2	1	6	5	9	8	3				1	8	4	3	9	6	2	7	5
5	8	3	2	9	4	6	7	1				7	5	9	1	8	2	4	6	3
2	1	5	9	3	7	4	6	8				5	1	3	7	4	9	6	8	2
3	6	4	8	5	2	1	9	7				8	4	7	6	2	1	3	5	9
8	7	9	6	4	1	3	5	2				2	9	6	8	3	5	7	4	1

59

1	4	6	2	7	3	5	9	8				5	2	8	6	7	4	1	3	9
9	8	3	1	5	6	2	4	7				6	3	1	5	2	9	4	7	8
5	7	2	4	8	9	3	1	6				9	4	7	3	1	8	2	5	6
6	5	4	8	9	1	7	2	3				8	7	9	2	5	3	6	1	4
2	1	7	6	3	5	4	8	9				4	6	2	7	8	1	3	9	5
8	3	9	7	2	4	6	5	1				1	5	3	9	4	6	8	2	7
4	9	1	5	6	7	8	3	2	5	6	1	7	9	4	8	3	2	5	6	1
3	6	8	9	4	2	1	7	5	4	2	9	3	8	6	1	9	5	7	4	2
7	2	5	3	1	8	9	6	4	3	8	7	2	1	5	4	6	7	9	8	3
						2	4	3	1	7	5	8	6	9						
						7	8	1	6	9	3	5	4	2						
						6	5	9	8	4	2	1	7	3						
4	2	3	7	1	8	5	9	6	7	3	8	4	2	1	7	8	9	6	3	5
1	9	7	6	4	5	3	2	8	9	1	4	6	5	7	4	3	2	8	9	1
6	8	5	3	2	9	4	1	7	2	5	6	9	3	8	1	6	5	2	4	7
9	3	6	1	8	7	2	5	4				7	6	9	5	2	8	4	1	3
8	4	2	5	9	6	7	3	1				2	4	3	6	1	7	9	5	8
5	7	1	4	3	2	8	6	9				8	1	5	3	9	4	7	2	6
7	5	4	2	6	1	9	8	3				3	7	6	9	4	1	5	8	2
3	6	9	8	5	4	1	7	2				5	9	2	8	7	3	1	6	4
2	1	8	9	7	3	6	4	5				1	8	4	2	5	6	3	7	9

60

7	5	1	8	4	9	3	6	2				2	7	5	4	3	9	6	1	8
4	6	2	5	7	3	8	9	1				4	8	6	5	1	2	9	3	7
8	3	9	2	1	6	5	7	4				1	9	3	8	6	7	5	2	4
3	4	7	1	6	8	9	2	5				9	3	7	2	5	6	8	4	1
2	8	5	7	9	4	6	1	3				6	4	8	3	7	1	2	9	5
9	1	6	3	2	5	7	4	8				5	2	1	9	4	8	3	7	6
5	7	4	6	3	2	1	8	9	4	7	6	3	5	2	7	8	4	1	6	9
1	2	3	9	8	7	4	5	6	8	2	3	7	1	9	6	2	5	4	8	3
6	9	8	4	5	1	2	3	7	1	5	9	8	6	4	1	9	3	7	5	2
						7	1	5	2	8	4	9	3	6						
						6	4	3	5	9	1	2	8	7						
						9	2	8	3	6	7	5	4	1						
3	7	1	2	9	5	8	6	4	9	3	2	1	7	5	3	8	6	4	2	9
6	2	5	7	8	4	3	9	1	7	4	5	6	2	8	1	9	4	7	5	3
9	8	4	1	6	3	5	7	2	6	1	8	4	9	3	7	5	2	1	8	6
4	1	9	5	7	6	2	8	3				5	1	6	2	3	8	9	4	7
7	5	3	9	2	8	1	4	6				9	4	2	5	6	7	3	1	8
8	6	2	3	4	1	7	5	9				8	3	7	9	4	1	5	6	2
5	9	6	8	1	2	4	3	7				3	6	1	8	7	5	2	9	4
1	3	7	4	5	9	6	2	8				7	5	4	6	2	9	8	3	1
2	4	8	6	3	7	9	1	5				2	8	9	4	1	3	6	7	5

SOLUTIONS

61

2	9	6	8	5	7	4	1	3				5	4	1	6	7	8	9	2	3
4	1	7	6	3	2	8	5	9				6	8	7	3	2	9	1	5	4
8	5	3	9	1	4	7	2	6				2	9	3	1	4	5	8	7	6
6	8	4	2	9	5	1	3	7				7	3	9	2	8	4	6	1	5
3	2	9	1	7	6	5	8	4				8	6	4	9	5	1	2	3	7
1	7	5	3	4	8	6	9	2				1	5	2	7	3	6	4	9	8
9	6	8	4	2	1	3	7	5	4	6	2	9	1	8	5	6	7	3	4	2
5	4	2	7	8	3	9	6	1	7	8	3	4	2	5	8	1	3	7	6	9
7	3	1	5	6	9	2	4	8	9	5	1	3	7	6	4	9	2	5	8	1
						4	5	9	6	1	7	2	8	3						
						6	1	2	8	3	5	7	9	4						
						8	3	7	2	4	9	6	5	1						
6	2	8	1	4	7	5	9	3	1	7	4	8	6	2	9	7	1	4	5	3
4	7	5	3	9	8	1	2	6	3	9	8	5	4	7	8	2	3	9	1	6
9	3	1	6	5	2	7	8	4	5	2	6	1	3	9	6	4	5	2	7	8
2	1	3	9	7	6	8	4	5				6	1	8	7	9	2	5	3	4
7	4	9	5	8	3	2	6	1				2	7	3	5	1	4	8	6	9
8	5	6	4	2	1	3	7	9				4	9	5	3	6	8	1	2	7
5	6	2	7	1	9	4	3	8				3	5	1	4	8	6	7	9	2
3	8	4	2	6	5	9	1	7				7	2	4	1	3	9	6	8	5
1	9	7	8	3	4	6	5	2				9	8	6	2	5	7	3	4	1

62

4	7	6	1	9	2	5	3	8				6	3	2	1	8	4	9	7	5
2	3	9	8	6	5	1	7	4				8	9	4	7	2	5	6	3	1
5	1	8	4	3	7	6	9	2				7	5	1	9	6	3	8	4	2
3	9	2	5	8	1	7	4	6				2	8	7	3	9	6	1	5	4
7	4	5	9	2	6	8	1	3				9	1	6	5	4	2	7	8	3
6	8	1	3	7	4	2	5	9				5	4	3	8	1	7	2	6	9
8	6	3	7	1	9	4	2	5	7	8	1	3	6	9	4	7	1	5	2	8
1	2	4	6	5	3	9	8	7	6	3	4	1	2	5	6	3	8	4	9	7
9	5	7	2	4	8	3	6	1	9	5	2	4	7	8	2	5	9	3	1	6
						1	9	2	3	7	8	5	4	6						
						6	4	8	2	9	5	7	1	3						
						7	5	3	1	4	6	9	8	2						
2	7	3	8	6	9	5	1	4	8	2	9	6	3	7	9	2	5	1	4	8
9	5	4	2	3	1	8	7	6	5	1	3	2	9	4	1	3	8	6	7	5
8	6	1	7	5	4	2	3	9	4	6	7	8	5	1	6	4	7	9	2	3
4	9	8	1	2	3	7	6	5				1	8	3	2	9	4	7	5	6
1	2	5	4	7	6	9	8	3				5	6	9	7	1	3	4	8	2
7	3	6	5	9	8	1	4	2				7	4	2	8	5	6	3	1	9
6	4	7	9	8	5	3	2	1				4	1	8	5	6	9	2	3	7
5	1	2	3	4	7	6	9	8				9	2	5	3	7	1	8	6	4
3	8	9	6	1	2	4	5	7				3	7	6	4	8	2	5	9	1

63

1	6	7	3	5	4	2	8	9				3	2	1	9	5	4	7	6	8
8	5	4	1	2	9	6	3	7				6	7	4	3	1	8	5	9	2
2	9	3	6	8	7	5	4	1				9	5	8	7	6	2	1	4	3
5	1	6	7	3	8	4	9	2				2	6	9	1	3	7	4	8	5
9	3	8	4	1	2	7	5	6				1	8	7	5	4	9	3	2	6
7	4	2	5	9	6	8	1	3				5	4	3	2	8	6	9	7	1
4	2	5	9	6	3	1	7	8	2	6	3	4	9	5	6	2	1	8	3	7
3	8	1	2	7	5	9	6	4	5	7	1	8	3	2	4	7	5	6	1	9
6	7	9	8	4	1	3	2	5	4	9	8	7	1	6	8	9	3	2	5	4
						6	9	3	7	1	5	2	4	8						
						5	8	2	3	4	6	1	7	9						
						7	4	1	8	2	9	5	6	3						
9	4	2	1	6	3	8	5	7	9	3	4	6	2	1	4	3	7	5	8	9
7	8	1	5	2	9	4	3	6	1	5	2	9	8	7	2	5	1	3	4	6
6	3	5	8	7	4	2	1	9	6	8	7	3	5	4	9	6	8	7	2	1
1	9	4	2	3	7	6	8	5				7	6	9	5	2	3	4	1	8
3	6	7	4	8	5	1	9	2				4	3	5	1	8	9	6	7	2
5	2	8	6	9	1	3	7	4				2	1	8	6	7	4	9	3	5
8	5	9	3	4	6	7	2	1				8	4	6	3	9	2	1	5	7
2	1	6	7	5	8	9	4	3				1	9	2	7	4	5	8	6	3
4	7	3	9	1	2	5	6	8				5	7	3	8	1	6	2	9	4

64

2	1	7	5	3	6	8	4	9				2	5	1	9	7	4	8	6	3
4	6	5	8	7	9	3	1	2				3	7	9	5	8	6	4	2	1
8	3	9	4	1	2	5	6	7				4	8	6	1	2	3	9	7	5
7	9	3	1	8	5	6	2	4				8	1	3	7	5	9	6	4	2
5	4	1	2	6	7	9	3	8				6	2	5	4	3	8	7	1	9
6	2	8	3	9	4	7	5	1				7	9	4	6	1	2	3	5	8
9	8	4	6	2	3	1	7	5	6	2	3	9	4	8	2	6	5	1	3	7
3	7	2	9	5	1	4	8	6	7	5	9	1	3	2	8	4	7	5	9	6
1	5	6	7	4	8	2	9	3	1	8	4	5	6	7	3	9	1	2	8	4
						6	2	8	5	4	1	7	9	3						
						9	1	4	3	7	6	2	8	5						
						3	5	7	2	9	8	4	1	6						
8	4	5	2	1	6	7	3	9	8	1	5	6	2	4	1	3	8	5	7	9
9	2	3	4	7	8	5	6	1	4	3	2	8	7	9	4	2	5	1	6	3
7	6	1	5	9	3	8	4	2	9	6	7	3	5	1	6	9	7	4	2	8
5	9	4	3	8	1	6	2	7				2	1	7	5	4	9	3	8	6
6	1	8	7	2	9	3	5	4				4	6	3	8	1	2	9	5	7
3	7	2	6	4	5	1	9	8				5	9	8	3	7	6	2	1	4
1	5	7	9	3	4	2	8	6				1	8	5	9	6	3	7	4	2
4	8	6	1	5	2	9	7	3				7	3	6	2	5	4	8	9	1
2	3	9	8	6	7	4	1	5				9	4	2	7	8	1	6	3	5

SOLUTIONS

65

5	1	4	9	6	8	3	7	2				3	1	4	8	7	9	6	5	2
7	3	8	1	2	5	4	9	6				9	2	6	5	3	1	7	4	8
9	6	2	7	4	3	1	8	5				7	8	5	2	6	4	9	1	3
4	2	5	6	9	1	8	3	7				6	9	3	4	5	7	2	8	1
3	7	9	5	8	2	6	4	1				8	5	2	9	1	6	3	7	4
1	8	6	3	7	4	2	5	9				4	7	1	3	2	8	5	6	9
2	4	7	8	1	9	5	6	3	7	8	1	2	4	9	7	8	5	1	3	6
6	5	1	4	3	7	9	2	8	4	6	5	1	3	7	6	4	2	8	9	5
8	9	3	2	5	6	7	1	4	9	3	2	5	6	8	1	9	3	4	2	7
						8	5	6	2	9	3	7	1	4						
						2	4	9	1	7	6	8	5	3						
						1	3	7	5	4	8	6	9	2						
5	7	3	4	2	8	6	9	1	3	2	7	4	8	5	3	6	9	7	2	1
8	2	4	9	6	1	3	7	5	8	1	4	9	2	6	4	1	7	3	8	5
1	9	6	5	3	7	4	8	2	6	5	9	3	7	1	8	5	2	6	4	9
3	1	9	2	4	6	7	5	8				7	1	3	6	9	4	8	5	2
6	5	7	3	8	9	1	2	4				6	5	2	1	7	8	4	9	3
4	8	2	7	1	5	9	3	6				8	4	9	2	3	5	1	6	7
7	4	5	6	9	2	8	1	3				2	6	7	9	4	1	5	3	8
9	3	8	1	5	4	2	6	7				1	3	8	5	2	6	9	7	4
2	6	1	8	7	3	5	4	9				5	9	4	7	8	3	2	1	6

66

3	9	5	8	1	4	2	6	7				7	1	6	8	2	9	4	3	5
6	7	2	9	5	3	1	8	4				3	8	9	5	7	4	2	1	6
4	8	1	6	7	2	5	3	9				5	4	2	3	1	6	9	7	8
7	4	6	1	3	9	8	5	2				4	2	1	6	3	8	5	9	7
1	5	8	4	2	6	7	9	3				9	3	8	4	5	7	1	6	2
9	2	3	7	8	5	4	1	6				6	7	5	2	9	1	3	8	4
2	3	7	5	6	8	9	4	1	8	7	5	2	6	3	9	8	5	7	4	1
8	1	4	3	9	7	6	2	5	4	1	3	8	9	7	1	4	2	6	5	3
5	6	9	2	4	1	3	7	8	6	9	2	1	5	4	7	6	3	8	2	9
						8	6	9	5	4	1	7	3	2						
						5	3	2	7	8	6	9	4	1						
						4	1	7	2	3	9	6	8	5						
6	2	7	4	8	9	1	5	3	9	2	8	4	7	6	1	9	2	5	3	8
4	3	9	5	2	1	7	8	6	1	5	4	3	2	9	5	6	8	4	7	1
8	5	1	7	6	3	2	9	4	3	6	7	5	1	8	7	4	3	2	6	9
3	9	2	8	1	5	4	6	7				2	6	1	8	5	7	3	9	4
7	4	8	6	9	2	5	3	1				8	4	3	2	1	9	6	5	7
1	6	5	3	4	7	8	2	9				9	5	7	6	3	4	1	8	2
2	7	6	9	5	4	3	1	8				6	9	2	4	8	5	7	1	3
9	1	3	2	7	8	6	4	5				7	3	5	9	2	1	8	4	6
5	8	4	1	3	6	9	7	2				1	8	4	3	7	6	9	2	5

67

4	5	2	6	7	8	1	3	9				5	3	7	1	2	8	9	6	4
8	1	9	5	4	3	7	6	2				4	6	8	3	7	9	5	2	1
6	7	3	2	1	9	5	4	8				9	2	1	4	6	5	8	3	7
5	8	7	1	9	4	6	2	3				3	7	9	8	4	1	6	5	2
2	9	6	3	8	5	4	7	1				1	8	2	6	5	7	4	9	3
1	3	4	7	6	2	8	9	5				6	4	5	9	3	2	1	7	8
9	4	1	8	2	6	3	5	7	8	4	1	2	9	6	7	8	4	3	1	5
7	2	5	4	3	1	9	8	6	3	5	2	7	1	4	5	9	3	2	8	6
3	6	8	9	5	7	2	1	4	9	6	7	8	5	3	2	1	6	7	4	9
						6	2	1	7	3	9	4	8	5						
						7	3	5	4	2	8	9	6	1						
						4	9	8	5	1	6	3	2	7						
1	4	9	7	3	5	8	6	2	1	7	3	5	4	9	3	7	8	2	1	6
6	8	3	1	2	4	5	7	9	6	8	4	1	3	2	9	5	6	4	8	7
7	5	2	9	6	8	1	4	3	2	9	5	6	7	8	2	1	4	5	3	9
8	6	7	2	5	9	4	3	1				7	1	5	6	4	9	8	2	3
2	9	1	4	7	3	6	5	8				4	9	3	8	2	5	7	6	1
4	3	5	6	8	1	2	9	7				8	2	6	1	3	7	9	4	5
3	2	4	8	9	6	7	1	5				3	8	1	5	9	2	6	7	4
9	1	8	5	4	7	3	2	6				9	6	7	4	8	1	3	5	2
5	7	6	3	1	2	9	8	4				2	5	4	7	6	3	1	9	8

68

9	7	8	3	1	4	5	6	2				7	2	3	4	5	8	6	1	9
4	1	2	9	5	6	3	7	8				4	6	8	2	9	1	3	7	5
5	3	6	8	7	2	9	4	1				5	9	1	6	3	7	8	2	4
7	5	9	2	4	8	1	3	6				6	1	5	9	2	4	7	3	8
1	2	3	7	6	9	8	5	4				8	3	4	5	7	6	2	9	1
8	6	4	5	3	1	7	2	9				2	7	9	1	8	3	5	4	6
3	9	1	6	2	5	4	8	7	2	9	3	1	5	6	3	4	2	9	8	7
2	8	5	4	9	7	6	1	3	4	5	7	9	8	2	7	1	5	4	6	3
6	4	7	1	8	3	2	9	5	8	1	6	3	4	7	8	6	9	1	5	2
						1	7	4	5	3	8	6	2	9						
						5	2	9	1	6	4	8	7	3						
						3	6	8	9	7	2	4	1	5						
5	8	1	6	7	4	9	3	2	7	4	1	5	6	8	7	4	3	2	9	1
2	7	9	1	5	3	8	4	6	3	2	5	7	9	1	8	2	6	3	4	5
6	4	3	9	8	2	7	5	1	6	8	9	2	3	4	1	9	5	6	8	7
1	5	7	4	2	6	3	9	8				8	5	7	6	1	4	9	2	3
3	2	6	8	9	5	4	1	7				4	2	9	5	3	7	1	6	8
8	9	4	7	3	1	2	6	5				6	1	3	2	8	9	5	7	4
4	3	5	2	1	8	6	7	9				9	4	2	3	5	8	7	1	6
9	1	2	3	6	7	5	8	4				1	7	5	4	6	2	8	3	9
7	6	8	5	4	9	1	2	3				3	8	6	9	7	1	4	5	2

SOLUTIONS

69

3	8	4	9	1	5	7	2	6				8	4	7	6	1	2	9	3	5
5	9	6	8	2	7	3	4	1				1	6	9	8	3	5	2	7	4
7	2	1	4	6	3	5	8	9				2	5	3	9	7	4	6	1	8
2	3	7	5	9	8	1	6	4				5	8	2	4	9	7	3	6	1
8	6	5	1	4	2	9	3	7				4	7	1	3	2	6	8	5	9
1	4	9	3	7	6	2	5	8				9	3	6	1	5	8	7	4	2
9	7	2	6	3	4	8	1	5	9	7	6	3	2	4	5	6	9	1	8	7
6	1	8	2	5	9	4	7	3	2	1	5	6	9	8	7	4	1	5	2	3
4	5	3	7	8	1	6	9	2	4	3	8	7	1	5	2	8	3	4	9	6
						9	5	4	7	8	2	1	3	6						
						2	8	6	3	5	1	9	4	7						
						1	3	7	6	9	4	5	8	2						
9	4	6	8	7	3	5	2	1	8	6	9	4	7	3	2	8	6	9	5	1
3	2	5	1	4	9	7	6	8	1	4	3	2	5	9	3	7	1	6	4	8
1	7	8	2	6	5	3	4	9	5	2	7	8	6	1	9	4	5	2	3	7
6	3	1	5	8	4	9	7	2				7	9	4	8	6	2	5	1	3
5	9	2	3	1	7	4	8	6				6	1	8	5	3	4	7	2	9
4	8	7	6	9	2	1	5	3				5	3	2	7	1	9	4	8	6
7	1	9	4	2	6	8	3	5				3	4	6	1	2	7	8	9	5
2	5	4	9	3	8	6	1	7				1	2	5	6	9	8	3	7	4
8	6	3	7	5	1	2	9	4				9	8	7	4	5	3	1	6	2

70

9	3	8	2	6	1	7	5	4				3	2	4	5	8	6	7	9	1
4	5	2	7	8	9	3	1	6				8	7	1	4	3	9	2	6	5
7	1	6	3	4	5	9	8	2				9	6	5	2	1	7	3	4	8
1	9	7	6	3	2	5	4	8				5	1	7	3	9	2	4	8	6
3	6	5	4	1	8	2	9	7				4	8	3	7	6	5	1	2	9
8	2	4	9	5	7	1	6	3				2	9	6	8	4	1	5	3	7
2	4	3	1	9	6	8	7	5	9	6	3	1	4	2	6	7	8	9	5	3
5	7	1	8	2	4	6	3	9	1	4	2	7	5	8	9	2	3	6	1	4
6	8	9	5	7	3	4	2	1	8	5	7	6	3	9	1	5	4	8	7	2
						3	8	4	5	7	1	2	9	6						
						9	1	6	4	2	8	3	7	5						
						7	5	2	6	3	9	4	8	1						
8	7	5	2	9	4	1	6	3	7	9	5	8	2	4	1	6	7	9	5	3
3	9	2	8	1	6	5	4	7	2	8	6	9	1	3	8	2	5	4	7	6
6	1	4	7	3	5	2	9	8	3	1	4	5	6	7	9	4	3	1	2	8
2	8	6	1	5	9	3	7	4				2	3	6	5	8	1	7	9	4
1	4	3	6	7	8	9	2	5				7	5	8	2	9	4	3	6	1
7	5	9	3	4	2	6	8	1				4	9	1	7	3	6	2	8	5
4	3	8	9	6	1	7	5	2				3	4	5	6	7	2	8	1	9
5	6	1	4	2	7	8	3	9				6	8	2	4	1	9	5	3	7
9	2	7	5	8	3	4	1	6				1	7	9	3	5	8	6	4	2

71

7	9	3	6	5	4	1	2	8				5	8	3	2	1	7	6	4	9
6	1	4	2	3	8	7	9	5				2	4	1	9	6	5	8	3	7
8	5	2	1	9	7	4	6	3				6	7	9	8	4	3	5	1	2
2	8	9	4	7	3	6	5	1				1	3	7	6	9	8	2	5	4
3	6	5	8	2	1	9	7	4				9	2	4	7	5	1	3	8	6
4	7	1	9	6	5	8	3	2				8	6	5	4	3	2	9	7	1
5	2	8	7	1	9	3	4	6	9	8	1	7	5	2	1	8	9	4	6	3
9	4	6	3	8	2	5	1	7	3	6	2	4	9	8	3	7	6	1	2	5
1	3	7	5	4	6	2	8	9	5	4	7	3	1	6	5	2	4	7	9	8
						1	5	2	8	7	3	6	4	9						
						7	6	4	2	1	9	8	3	5						
						8	9	3	4	5	6	1	2	7						
8	4	1	3	7	9	6	2	5	1	3	8	9	7	4	1	3	2	8	6	5
9	2	7	6	5	1	4	3	8	7	9	5	2	6	1	5	4	8	9	7	3
5	3	6	4	8	2	9	7	1	6	2	4	5	8	3	7	9	6	1	2	4
7	6	4	2	3	8	1	5	9				7	9	5	4	6	1	2	3	8
2	1	5	9	4	7	3	8	6				3	1	8	9	2	7	4	5	6
3	8	9	1	6	5	2	4	7				6	4	2	8	5	3	7	1	9
1	7	3	5	2	6	8	9	4				1	2	9	6	8	5	3	4	7
4	9	8	7	1	3	5	6	2				4	3	6	2	7	9	5	8	1
6	5	2	8	9	4	7	1	3				8	5	7	3	1	4	6	9	2

72

7	2	3	4	6	5	1	8	9				5	2	3	7	6	9	1	8	4
5	8	9	2	7	1	4	6	3				6	9	1	8	5	4	2	3	7
6	1	4	8	9	3	2	5	7				8	4	7	1	2	3	5	9	6
9	5	6	7	8	2	3	4	1				4	6	9	2	3	1	7	5	8
1	7	8	6	3	4	5	9	2				3	7	2	5	4	8	6	1	9
3	4	2	1	5	9	8	7	6				1	5	8	6	9	7	3	4	2
8	3	5	9	1	6	7	2	4	8	6	1	9	3	5	4	7	6	8	2	1
4	6	1	5	2	7	9	3	8	5	7	4	2	1	6	9	8	5	4	7	3
2	9	7	3	4	8	6	1	5	2	9	3	7	8	4	3	1	2	9	6	5
						3	7	2	6	8	9	4	5	1						
						4	6	9	1	3	5	8	7	2						
						8	5	1	4	2	7	6	9	3						
1	2	4	9	6	7	5	8	3	7	4	6	1	2	9	8	6	7	4	5	3
6	9	8	5	3	2	1	4	7	9	5	2	3	6	8	4	5	2	9	7	1
5	7	3	4	8	1	2	9	6	3	1	8	5	4	7	9	3	1	6	8	2
9	6	7	8	2	4	3	5	1				9	5	1	3	8	4	7	2	6
8	1	2	3	5	9	6	7	4				4	8	2	7	1	6	5	3	9
3	4	5	1	7	6	8	2	9				6	7	3	5	2	9	8	1	4
7	3	1	2	9	8	4	6	5				8	9	4	2	7	3	1	6	5
4	8	6	7	1	5	9	3	2				7	3	6	1	9	5	2	4	8
2	5	9	6	4	3	7	1	8				2	1	5	6	4	8	3	9	7

SOLUTIONS

73

2	8	4	7	6	3	9	5	1				4	7	1	3	5	8	2	9	6
5	7	1	4	2	9	6	8	3				8	3	9	2	1	6	4	5	7
6	3	9	1	8	5	2	7	4				6	5	2	9	4	7	3	8	1
8	2	6	5	3	1	4	9	7				9	2	7	8	3	5	1	6	4
3	9	5	6	7	4	1	2	8				3	6	8	1	9	4	7	2	5
1	4	7	2	9	8	3	6	5				1	4	5	7	6	2	9	3	8
9	6	3	8	4	7	5	1	2	8	4	6	7	9	3	6	8	1	5	4	2
7	5	2	3	1	6	8	4	9	2	3	7	5	1	6	4	2	9	8	7	3
4	1	8	9	5	2	7	3	6	5	1	9	2	8	4	5	7	3	6	1	9
						6	7	5	9	8	3	1	4	2						
						4	9	3	7	2	1	6	5	8						
						1	2	8	6	5	4	9	3	7						
5	9	6	7	3	4	2	8	1	3	7	5	4	6	9	7	8	2	5	3	1
8	1	7	5	2	9	3	6	4	1	9	2	8	7	5	1	3	4	6	9	2
2	4	3	6	8	1	9	5	7	4	6	8	3	2	1	5	9	6	8	7	4
4	2	1	8	9	6	5	7	3				2	5	8	4	6	7	3	1	9
3	7	8	2	1	5	4	9	6				9	3	7	2	1	8	4	6	5
9	6	5	3	4	7	1	2	8				6	1	4	9	5	3	7	2	8
1	5	4	9	7	8	6	3	2				7	9	6	8	4	1	2	5	3
7	3	9	1	6	2	8	4	5				5	4	3	6	2	9	1	8	7
6	8	2	4	5	3	7	1	9				1	8	2	3	7	5	9	4	6

74

9	5	6	7	1	8	3	4	2				1	9	7	4	8	2	5	6	3
2	4	1	6	5	3	9	8	7				2	6	8	3	5	9	4	7	1
3	8	7	9	4	2	5	6	1				4	5	3	7	6	1	9	2	8
8	2	9	3	6	5	1	7	4				7	1	6	2	4	8	3	5	9
4	6	3	8	7	1	2	5	9				8	2	5	9	3	7	6	1	4
1	7	5	2	9	4	6	3	8				3	4	9	6	1	5	2	8	7
6	9	2	5	8	7	4	1	3	7	6	9	5	8	2	1	9	4	7	3	6
5	1	8	4	3	9	7	2	6	4	8	5	9	3	1	5	7	6	8	4	2
7	3	4	1	2	6	8	9	5	2	3	1	6	7	4	8	2	3	1	9	5
						9	8	1	6	2	4	3	5	7						
						3	7	2	9	5	8	4	1	6						
						5	6	4	3	1	7	8	2	9						
9	5	8	6	3	1	2	4	7	5	9	3	1	6	8	2	3	7	9	4	5
1	2	4	7	9	5	6	3	8	1	4	2	7	9	5	6	1	4	8	3	2
3	6	7	4	2	8	1	5	9	8	7	6	2	4	3	5	9	8	7	6	1
8	9	2	5	4	3	7	6	1				5	7	6	1	2	3	4	8	9
6	3	5	2	1	7	8	9	4				8	1	9	4	7	5	6	2	3
7	4	1	8	6	9	3	2	5				4	3	2	8	6	9	5	1	7
5	1	3	9	7	6	4	8	2				6	2	4	9	5	1	3	7	8
2	7	9	3	8	4	5	1	6				3	5	1	7	8	6	2	9	4
4	8	6	1	5	2	9	7	3				9	8	7	3	4	2	1	5	6

75

7	2	8	6	9	4	5	1	3				8	9	6	7	5	1	4	3	2
6	9	5	3	8	1	7	4	2				1	5	3	2	4	6	9	8	7
4	3	1	7	2	5	6	8	9				7	2	4	3	8	9	5	1	6
2	1	4	5	6	8	9	3	7				4	8	2	1	7	5	3	6	9
8	6	3	4	7	9	2	5	1				3	1	5	6	9	2	8	7	4
9	5	7	1	3	2	8	6	4				6	7	9	8	3	4	1	2	5
5	4	9	2	1	6	3	7	8	6	9	5	2	4	1	9	6	3	7	5	8
3	8	6	9	4	7	1	2	5	7	4	3	9	6	8	5	1	7	2	4	3
1	7	2	8	5	3	4	9	6	2	8	1	5	3	7	4	2	8	6	9	1
						6	4	9	8	5	2	7	1	3						
						8	1	7	4	3	9	6	5	2						
						5	3	2	1	6	7	8	9	4						
3	2	5	9	6	1	7	8	4	5	1	6	3	2	9	8	1	4	5	6	7
8	1	9	5	4	7	2	6	3	9	7	4	1	8	5	7	6	9	2	4	3
4	7	6	3	8	2	9	5	1	3	2	8	4	7	6	2	5	3	8	1	9
7	8	1	4	2	3	5	9	6				5	9	3	4	8	2	1	7	6
2	5	3	6	1	9	4	7	8				2	1	8	6	9	7	4	3	5
9	6	4	8	7	5	1	3	2				6	4	7	1	3	5	9	8	2
5	4	2	7	3	8	6	1	9				8	3	2	9	7	1	6	5	4
6	3	7	1	9	4	8	2	5				9	5	1	3	4	6	7	2	8
1	9	8	2	5	6	3	4	7				7	6	4	5	2	8	3	9	1

76

4	3	1	9	2	7	6	8	5				4	6	9	3	2	7	1	8	5
9	2	6	8	3	5	1	7	4				8	5	1	4	6	9	3	7	2
5	7	8	4	1	6	9	3	2				7	3	2	8	1	5	6	4	9
6	9	3	7	4	8	2	5	1				2	4	5	9	7	6	8	3	1
8	5	2	6	9	1	3	4	7				1	8	6	2	4	3	9	5	7
1	4	7	3	5	2	8	9	6				3	9	7	1	5	8	2	6	4
2	1	9	5	8	4	7	6	3	1	9	8	5	2	4	6	3	1	7	9	8
7	8	5	2	6	3	4	1	9	2	3	5	6	7	8	5	9	2	4	1	3
3	6	4	1	7	9	5	2	8	7	4	6	9	1	3	7	8	4	5	2	6
						8	5	1	4	7	3	2	6	9						
						2	4	6	8	1	9	3	5	7						
						9	3	7	5	6	2	8	4	1						
6	2	9	7	4	3	1	8	5	3	2	7	4	9	6	5	2	3	8	7	1
8	1	7	5	2	6	3	9	4	6	5	1	7	8	2	1	4	9	6	5	3
5	4	3	9	8	1	6	7	2	9	8	4	1	3	5	6	7	8	4	9	2
3	9	1	6	7	2	5	4	8				2	6	1	3	8	7	5	4	9
2	6	8	1	5	4	7	3	9				8	4	9	2	5	6	1	3	7
7	5	4	3	9	8	2	1	6				3	5	7	4	9	1	2	6	8
1	8	5	2	3	9	4	6	7				9	2	4	8	3	5	7	1	6
4	3	2	8	6	7	9	5	1				6	7	8	9	1	4	3	2	5
9	7	6	4	1	5	8	2	3				5	1	3	7	6	2	9	8	4

SOLUTIONS

77

6	2	7	8	3	5	1	9	4				8	9	3	7	1	5	4	6	2
9	4	8	2	6	1	7	5	3				7	4	2	6	8	9	3	1	5
1	3	5	9	7	4	2	8	6				1	6	5	4	2	3	9	8	7
2	9	4	1	8	6	5	3	7				2	5	9	3	7	8	1	4	6
5	7	6	4	2	3	9	1	8				6	7	4	2	9	1	8	5	3
8	1	3	5	9	7	4	6	2				3	8	1	5	4	6	7	2	9
3	8	2	7	1	9	6	4	5	1	3	8	9	2	7	1	6	4	5	3	8
7	5	1	6	4	8	3	2	9	4	6	7	5	1	8	9	3	2	6	7	4
4	6	9	3	5	2	8	7	1	2	9	5	4	3	6	8	5	7	2	9	1
						1	6	7	5	4	9	3	8	2						
						4	8	3	7	1	2	6	5	9						
						9	5	2	6	8	3	1	7	4						
4	5	7	9	6	1	2	3	8	9	5	4	7	6	1	3	4	5	2	8	9
6	8	9	2	3	5	7	1	4	3	2	6	8	9	5	2	1	7	6	3	4
1	3	2	8	4	7	5	9	6	8	7	1	2	4	3	8	9	6	5	7	1
5	2	1	3	9	6	4	8	7				4	3	2	6	5	9	7	1	8
3	6	4	5	7	8	9	2	1				1	7	6	4	3	8	9	5	2
7	9	8	4	1	2	6	5	3				9	5	8	7	2	1	3	4	6
8	7	6	1	2	9	3	4	5				3	8	9	5	6	4	1	2	7
2	1	3	6	5	4	8	7	9				5	1	4	9	7	2	8	6	3
9	4	5	7	8	3	1	6	2				6	2	7	1	8	3	4	9	5

78

7	8	9	4	2	5	3	6	1				2	4	1	7	6	8	9	5	3
1	2	3	7	6	8	9	4	5				8	7	3	1	5	9	4	2	6
6	4	5	1	3	9	7	2	8				6	9	5	4	2	3	1	8	7
4	3	1	9	8	7	2	5	6				3	6	7	8	9	1	5	4	2
8	6	7	5	1	2	4	3	9				5	2	8	6	4	7	3	1	9
5	9	2	6	4	3	8	1	7				9	1	4	2	3	5	7	6	8
3	1	4	8	9	6	5	7	2	6	4	8	1	3	9	5	8	6	2	7	4
2	7	8	3	5	1	6	9	4	5	1	3	7	8	2	3	1	4	6	9	5
9	5	6	2	7	4	1	8	3	2	7	9	4	5	6	9	7	2	8	3	1
						7	2	9	1	8	6	3	4	5						
						3	1	8	4	5	2	9	6	7						
						4	5	6	3	9	7	2	1	8						
4	5	7	8	9	6	2	3	1	7	6	5	8	9	4	1	5	6	2	7	3
6	2	8	3	1	7	9	4	5	8	2	1	6	7	3	4	8	2	9	5	1
3	1	9	5	4	2	8	6	7	9	3	4	5	2	1	3	9	7	4	6	8
2	4	3	1	7	8	6	5	9				3	6	2	9	4	1	7	8	5
5	8	1	9	6	4	7	2	3				9	1	5	8	7	3	6	2	4
9	7	6	2	3	5	1	8	4				7	4	8	6	2	5	1	3	9
1	6	2	4	5	9	3	7	8				4	5	9	7	6	8	3	1	2
8	3	5	7	2	1	4	9	6				2	3	6	5	1	9	8	4	7
7	9	4	6	8	3	5	1	2				1	8	7	2	3	4	5	9	6

79

3	8	7	4	5	9	1	2	6				4	1	3	8	5	9	7	2	6
4	2	5	8	6	1	9	7	3				7	9	8	6	2	1	3	5	4
1	6	9	2	3	7	5	4	8				6	2	5	7	4	3	8	1	9
9	4	6	3	8	2	7	1	5				1	6	9	2	3	8	5	4	7
5	3	8	1	7	4	2	6	9				2	8	4	5	6	7	9	3	1
2	7	1	5	9	6	3	8	4				3	5	7	9	1	4	2	6	8
6	1	3	9	2	8	4	5	7	2	9	1	8	3	6	1	9	5	4	7	2
8	5	2	7	4	3	6	9	1	4	8	3	5	7	2	4	8	6	1	9	3
7	9	4	6	1	5	8	3	2	5	6	7	9	4	1	3	7	2	6	8	5
						2	4	3	7	5	9	6	1	8						
						7	6	9	8	1	2	4	5	3						
						1	8	5	3	4	6	2	9	7						
7	4	6	3	5	2	9	1	8	6	3	5	7	2	4	8	6	9	5	1	3
9	8	3	7	6	1	5	2	4	1	7	8	3	6	9	7	1	5	4	8	2
1	5	2	8	9	4	3	7	6	9	2	4	1	8	5	3	2	4	9	7	6
6	1	4	5	3	9	7	8	2				8	4	6	9	7	2	3	5	1
3	7	9	1	2	8	4	6	5				2	7	3	5	8	1	6	9	4
8	2	5	4	7	6	1	9	3				9	5	1	4	3	6	7	2	8
4	9	8	2	1	3	6	5	7				5	3	2	6	9	8	1	4	7
5	3	1	6	8	7	2	4	9				4	1	7	2	5	3	8	6	9
2	6	7	9	4	5	8	3	1				6	9	8	1	4	7	2	3	5

80

4	1	8	2	7	6	5	9	3				6	5	3	2	8	9	7	4	1
2	9	7	3	4	5	8	1	6				7	2	8	5	1	4	9	6	3
3	6	5	9	1	8	2	4	7				4	9	1	7	3	6	2	8	5
5	7	3	1	8	4	6	2	9				9	4	7	6	2	5	1	3	8
8	2	6	7	9	3	1	5	4				3	1	5	8	9	7	4	2	6
1	4	9	5	6	2	3	7	8				8	6	2	1	4	3	5	7	9
7	3	1	8	2	9	4	6	5	1	8	3	2	7	9	3	6	1	8	5	4
6	5	2	4	3	7	9	8	1	7	6	2	5	3	4	9	7	8	6	1	2
9	8	4	6	5	1	7	3	2	5	4	9	1	8	6	4	5	2	3	9	7
						8	7	6	9	2	1	4	5	3						
						1	5	4	6	3	7	9	2	8						
						3	2	9	4	5	8	7	6	1						
3	1	5	4	7	2	6	9	8	2	1	5	3	4	7	5	9	8	6	2	1
4	9	7	6	8	5	2	1	3	8	7	4	6	9	5	4	2	1	3	7	8
6	8	2	1	3	9	5	4	7	3	9	6	8	1	2	6	7	3	4	9	5
9	4	8	5	2	7	3	6	1				1	3	4	2	5	6	9	8	7
5	7	3	8	1	6	9	2	4				7	2	9	8	1	4	5	3	6
1	2	6	9	4	3	8	7	5				5	6	8	7	3	9	1	4	2
7	6	9	3	5	4	1	8	2				2	7	3	1	4	5	8	6	9
8	5	4	2	6	1	7	3	9				9	8	1	3	6	7	2	5	4
2	3	1	7	9	8	4	5	6				4	5	6	9	8	2	7	1	3

SOLUTIONS

81

4	3	5	9	6	7	2	8	1				5	3	9	8	4	6	1	7	2
6	9	2	5	1	8	3	4	7				2	6	8	3	7	1	5	4	9
8	1	7	2	4	3	6	5	9				4	1	7	9	2	5	3	6	8
2	5	3	8	9	1	4	7	6				7	4	2	5	1	8	9	3	6
9	4	1	6	7	5	8	3	2				8	9	3	2	6	7	4	5	1
7	8	6	3	2	4	9	1	5				1	5	6	4	3	9	2	8	7
1	6	8	7	3	2	5	9	4	6	2	7	3	8	1	6	5	2	7	9	4
3	2	4	1	5	9	7	6	8	5	1	3	9	2	4	7	8	3	6	1	5
5	7	9	4	8	6	1	2	3	4	9	8	6	7	5	1	9	4	8	2	3
						2	1	6	8	4	5	7	3	9						
						4	5	9	7	3	2	1	6	8						
						8	3	7	9	6	1	5	4	2						
1	3	4	7	5	6	9	8	2	3	5	6	4	1	7	8	6	9	5	2	3
8	7	5	9	3	2	6	4	1	2	7	9	8	5	3	4	7	2	9	1	6
2	9	6	4	1	8	3	7	5	1	8	4	2	9	6	5	3	1	8	7	4
4	1	7	3	2	9	8	5	6				5	6	8	2	1	7	4	3	9
3	6	9	8	7	5	1	2	4				7	4	9	3	5	8	1	6	2
5	2	8	1	6	4	7	3	9				1	3	2	6	9	4	7	8	5
6	5	1	2	8	7	4	9	3				9	2	5	1	8	3	6	4	7
9	8	3	5	4	1	2	6	7				3	7	1	9	4	6	2	5	8
7	4	2	6	9	3	5	1	8				6	8	4	7	2	5	3	9	1

82

7	5	1	8	4	9	3	6	2				2	7	5	4	3	9	6	1	8
4	6	2	5	7	3	8	9	1				4	8	6	5	1	2	9	3	7
8	3	9	2	1	6	5	7	4				1	9	3	8	6	7	5	2	4
3	4	7	1	6	8	9	2	5				9	3	7	2	5	6	8	4	1
2	8	5	7	9	4	6	1	3				6	4	8	3	7	1	2	9	5
9	1	6	3	2	5	7	4	8				5	2	1	9	4	8	3	7	6
5	7	4	6	3	2	1	8	9	4	7	6	3	5	2	7	8	4	1	6	9
1	2	3	9	8	7	4	5	6	8	2	3	7	1	9	6	2	5	4	8	3
6	9	8	4	5	1	2	3	7	1	5	9	8	6	4	1	9	3	7	5	2
						7	1	5	2	8	4	9	3	6						
						6	4	3	5	9	1	2	8	7						
						9	2	8	3	6	7	5	4	1						
3	7	1	2	9	5	8	6	4	9	3	2	1	7	5	3	8	6	4	2	9
6	2	5	7	8	4	3	9	1	7	4	5	6	2	8	1	9	4	7	5	3
9	8	4	1	6	3	5	7	2	6	1	8	4	9	3	7	5	2	1	8	6
4	1	9	5	7	6	2	8	3				5	1	6	2	3	8	9	4	7
7	5	3	9	2	8	1	4	6				9	4	2	5	6	7	3	1	8
8	6	2	3	4	1	7	5	9				8	3	7	9	4	1	5	6	2
5	9	6	8	1	2	4	3	7				3	6	1	8	7	5	2	9	4
1	3	7	4	5	9	6	2	8				7	5	4	6	2	9	8	3	1
2	4	8	6	3	7	9	1	5				2	8	9	4	1	3	6	7	5

83

2	6	4	7	5	1	9	3	8				3	5	2	4	9	7	6	8	1
3	8	7	6	9	2	4	1	5				9	6	1	3	2	8	5	4	7
5	1	9	3	4	8	7	6	2				8	7	4	1	6	5	3	9	2
6	2	5	8	3	4	1	7	9				2	1	5	8	7	6	9	3	4
7	4	8	9	1	6	2	5	3				4	8	6	9	1	3	2	7	5
1	9	3	5	2	7	6	8	4				7	9	3	2	5	4	1	6	8
8	5	1	4	6	9	3	2	7	8	1	6	5	4	9	7	3	1	8	2	6
9	7	2	1	8	3	5	4	6	2	7	9	1	3	8	6	4	2	7	5	9
4	3	6	2	7	5	8	9	1	4	3	5	6	2	7	5	8	9	4	1	3
						2	3	8	6	5	1	9	7	4						
						7	6	5	3	9	4	8	1	2						
						4	1	9	7	2	8	3	5	6						
5	7	1	9	4	2	6	8	3	1	4	7	2	9	5	7	3	8	4	6	1
6	3	4	5	8	1	9	7	2	5	6	3	4	8	1	6	5	9	7	3	2
2	8	9	6	7	3	1	5	4	9	8	2	7	6	3	4	2	1	5	8	9
9	1	6	4	2	8	7	3	5				5	2	6	1	9	3	8	4	7
7	5	8	3	9	6	4	2	1				8	4	7	2	6	5	9	1	3
4	2	3	1	5	7	8	9	6				3	1	9	8	7	4	6	2	5
3	4	7	2	1	9	5	6	8				1	3	8	5	4	7	2	9	6
8	6	5	7	3	4	2	1	9				9	7	2	3	8	6	1	5	4
1	9	2	8	6	5	3	4	7				6	5	4	9	1	2	3	7	8

84

1	4	9	7	5	3	6	8	2				5	4	9	1	3	7	2	8	6
5	7	6	2	8	4	1	9	3				1	8	3	9	2	6	4	7	5
8	3	2	9	1	6	7	4	5				6	2	7	4	8	5	3	9	1
9	2	4	6	7	5	8	3	1				3	5	1	8	7	9	6	4	2
3	8	7	4	2	1	9	5	6				7	9	4	6	5	2	1	3	8
6	1	5	3	9	8	2	7	4				2	6	8	3	4	1	9	5	7
2	5	1	8	4	7	3	6	9	7	2	8	4	1	5	2	9	8	7	6	3
4	6	8	1	3	9	5	2	7	1	9	4	8	3	6	7	1	4	5	2	9
7	9	3	5	6	2	4	1	8	6	5	3	9	7	2	5	6	3	8	1	4
						2	7	1	4	8	5	6	9	3						
						9	8	4	3	1	6	5	2	7						
						6	5	3	2	7	9	1	4	8						
7	9	4	1	6	2	8	3	5	9	4	2	7	6	1	2	8	3	9	5	4
6	8	3	4	5	7	1	9	2	8	6	7	3	5	4	9	6	7	2	1	8
1	5	2	8	3	9	7	4	6	5	3	1	2	8	9	1	4	5	6	3	7
9	4	6	5	2	1	3	8	7				6	1	8	3	5	4	7	2	9
5	7	8	3	9	4	2	6	1				9	7	5	6	1	2	8	4	3
3	2	1	6	7	8	4	5	9				4	2	3	8	7	9	1	6	5
4	1	5	2	8	6	9	7	3				1	9	2	4	3	8	5	7	6
8	3	7	9	1	5	6	2	4				8	3	7	5	2	6	4	9	1
2	6	9	7	4	3	5	1	8				5	4	6	7	9	1	3	8	2

SOLUTIONS

85

6	5	2	1	7	9	8	3	4				5	4	7	6	8	9	2	1	3
1	9	8	5	4	3	2	6	7				8	1	2	5	4	3	6	9	7
3	7	4	2	8	6	9	5	1				3	9	6	7	1	2	5	8	4
8	3	1	4	6	5	7	9	2				1	3	8	2	7	6	4	5	9
2	4	5	3	9	7	1	8	6				7	2	4	8	9	5	3	6	1
9	6	7	8	1	2	5	4	3				9	6	5	4	3	1	8	7	2
5	2	9	7	3	4	6	1	8	5	9	2	4	7	3	1	6	8	9	2	5
4	8	6	9	2	1	3	7	5	6	1	4	2	8	9	3	5	7	1	4	6
7	1	3	6	5	8	4	2	9	3	7	8	6	5	1	9	2	4	7	3	8
						2	4	1	8	5	3	9	6	7						
						7	8	6	1	2	9	3	4	5						
						5	9	3	4	6	7	8	1	2						
3	6	1	9	7	2	8	5	4	9	3	1	7	2	6	4	3	8	1	5	9
9	2	8	6	4	5	1	3	7	2	4	6	5	9	8	1	7	6	3	4	2
7	5	4	8	1	3	9	6	2	7	8	5	1	3	4	9	2	5	8	7	6
4	1	6	5	3	7	2	9	8				9	7	5	3	1	2	6	8	4
8	7	5	2	6	9	4	1	3				6	4	2	7	8	9	5	1	3
2	9	3	1	8	4	5	7	6				3	8	1	6	5	4	2	9	7
1	3	7	4	5	8	6	2	9				4	5	3	2	9	1	7	6	8
6	8	2	7	9	1	3	4	5				8	6	7	5	4	3	9	2	1
5	4	9	3	2	6	7	8	1				2	1	9	8	6	7	4	3	5

86

3	7	1	6	4	5	8	9	2				6	1	9	4	3	2	5	8	7
4	2	5	3	9	8	6	1	7				5	8	7	1	6	9	3	4	2
9	8	6	1	7	2	5	3	4				2	4	3	7	5	8	1	6	9
5	4	2	7	8	3	9	6	1				4	9	5	2	1	6	8	7	3
7	1	9	4	5	6	3	2	8				3	7	1	5	8	4	2	9	6
8	6	3	2	1	9	4	7	5				8	6	2	3	9	7	4	5	1
1	9	7	5	3	4	2	8	6	9	3	7	1	5	4	9	7	3	6	2	8
2	5	8	9	6	1	7	4	3	1	6	5	9	2	8	6	4	1	7	3	5
6	3	4	8	2	7	1	5	9	2	8	4	7	3	6	8	2	5	9	1	4
						9	2	7	5	4	1	8	6	3						
						8	1	4	3	7	6	5	9	2						
						6	3	5	8	2	9	4	7	1						
4	6	9	5	8	1	3	7	2	4	5	8	6	1	9	4	3	5	7	2	8
7	1	2	4	6	3	5	9	8	6	1	3	2	4	7	8	9	1	5	6	3
5	8	3	2	9	7	4	6	1	7	9	2	3	8	5	7	2	6	4	9	1
9	2	6	3	7	4	8	1	5				8	3	4	1	7	9	6	5	2
3	5	1	8	2	6	9	4	7				1	9	6	5	4	2	8	3	7
8	4	7	9	1	5	2	3	6				5	7	2	6	8	3	1	4	9
6	9	4	1	5	2	7	8	3				4	6	3	9	1	8	2	7	5
1	3	5	7	4	8	6	2	9				9	5	8	2	6	7	3	1	4
2	7	8	6	3	9	1	5	4				7	2	1	3	5	4	9	8	6

87

7	6	4	5	8	3	9	2	1				2	5	7	9	1	8	6	3	4
9	1	5	7	4	2	6	8	3				4	8	9	3	7	6	2	1	5
3	2	8	6	1	9	7	4	5				3	6	1	5	2	4	7	9	8
5	9	6	3	2	4	8	1	7				7	2	4	1	5	3	9	8	6
1	4	7	8	9	6	5	3	2				8	9	3	4	6	7	1	5	2
2	8	3	1	5	7	4	6	9				6	1	5	2	8	9	4	7	3
4	5	9	2	6	1	3	7	8	9	1	6	5	4	2	7	3	1	8	6	9
6	3	2	9	7	8	1	5	4	3	8	2	9	7	6	8	4	5	3	2	1
8	7	1	4	3	5	2	9	6	5	7	4	1	3	8	6	9	2	5	4	7
						4	1	9	8	2	3	7	6	5						
						5	3	7	1	6	9	2	8	4						
						8	6	2	7	4	5	3	9	1						
9	6	3	1	8	4	7	2	5	6	3	8	4	1	9	2	3	7	6	8	5
4	1	2	5	7	6	9	8	3	4	5	1	6	2	7	9	5	8	4	1	3
7	8	5	9	2	3	6	4	1	2	9	7	8	5	3	4	1	6	7	2	9
5	9	6	3	1	8	2	7	4				2	8	6	7	9	1	3	5	4
1	4	7	2	6	5	3	9	8				7	3	5	8	2	4	9	6	1
2	3	8	7	4	9	1	5	6				9	4	1	5	6	3	8	7	2
3	2	4	6	5	7	8	1	9				1	6	8	3	4	5	2	9	7
6	5	1	8	9	2	4	3	7				3	7	2	1	8	9	5	4	6
8	7	9	4	3	1	5	6	2				5	9	4	6	7	2	1	3	8

88

8	2	9	5	7	1	6	3	4				5	3	9	6	8	1	2	4	7
7	6	3	2	4	8	9	5	1				6	8	2	7	5	4	3	9	1
4	5	1	3	6	9	7	8	2				1	7	4	9	2	3	6	5	8
2	8	6	9	5	3	1	4	7				8	5	7	4	3	9	1	2	6
1	3	4	6	2	7	5	9	8				9	4	1	2	7	6	5	8	3
5	9	7	8	1	4	2	6	3				2	6	3	5	1	8	4	7	9
3	1	8	7	9	6	4	2	5	1	7	8	3	9	6	8	4	2	7	1	5
6	7	5	4	8	2	3	1	9	4	6	5	7	2	8	1	6	5	9	3	4
9	4	2	1	3	5	8	7	6	3	2	9	4	1	5	3	9	7	8	6	2
						6	5	1	9	4	3	8	7	2						
						7	4	2	5	8	6	1	3	9						
						9	3	8	7	1	2	6	5	4						
6	1	5	8	7	3	2	9	4	8	3	7	5	6	1	2	9	4	3	8	7
4	7	9	1	2	6	5	8	3	6	9	1	2	4	7	8	1	3	5	9	6
8	3	2	4	9	5	1	6	7	2	5	4	9	8	3	6	5	7	4	1	2
2	5	1	6	4	9	3	7	8				6	2	8	1	3	9	7	5	4
3	6	4	7	5	8	9	2	1				3	7	9	4	6	5	1	2	8
7	9	8	2	3	1	6	4	5				4	1	5	7	8	2	6	3	9
9	4	7	3	1	2	8	5	6				8	9	6	5	7	1	2	4	3
1	2	6	5	8	4	7	3	9				1	3	2	9	4	6	8	7	5
5	8	3	9	6	7	4	1	2				7	5	4	3	2	8	9	6	1

SOLUTIONS

89

2	3	7	5	9	4	6	1	8				6	1	7	4	5	2	3	9	8
4	6	8	3	1	7	5	2	9				5	4	8	1	3	9	2	6	7
9	1	5	8	6	2	4	3	7				2	9	3	6	8	7	4	1	5
3	9	1	7	8	5	2	6	4				4	2	6	8	1	3	7	5	9
7	4	2	9	3	6	8	5	1				7	8	1	5	9	4	6	2	3
5	8	6	4	2	1	7	9	3				9	3	5	7	2	6	8	4	1
6	5	4	1	7	9	3	8	2	6	7	9	1	5	4	3	6	8	9	7	2
8	7	9	2	5	3	1	4	6	8	2	5	3	7	9	2	4	1	5	8	6
1	2	3	6	4	8	9	7	5	1	4	3	8	6	2	9	7	5	1	3	4
						5	2	9	7	3	1	6	4	8						
						8	3	4	2	9	6	7	1	5						
						6	1	7	4	5	8	9	2	3						
6	3	1	9	2	4	7	5	8	3	1	2	4	9	6	3	5	1	2	7	8
8	4	9	3	5	7	2	6	1	9	8	4	5	3	7	6	2	8	4	9	1
2	7	5	1	6	8	4	9	3	5	6	7	2	8	1	7	9	4	3	5	6
4	2	6	5	7	1	3	8	9				9	1	4	5	6	3	8	2	7
9	5	7	4	8	3	6	1	2				3	6	2	9	8	7	1	4	5
3	1	8	6	9	2	5	7	4				7	5	8	4	1	2	6	3	9
7	8	4	2	1	5	9	3	6				1	7	5	2	4	6	9	8	3
1	6	3	7	4	9	8	2	5				6	4	9	8	3	5	7	1	2
5	9	2	8	3	6	1	4	7				8	2	3	1	7	9	5	6	4

90

5	6	1	2	9	7	4	8	3				8	7	3	1	9	5	6	2	4
7	4	9	3	8	5	1	2	6				2	6	9	3	8	4	1	7	5
8	3	2	6	1	4	5	9	7				1	4	5	2	7	6	3	9	8
9	7	4	5	2	8	3	6	1				7	9	2	6	5	8	4	1	3
6	1	8	4	7	3	9	5	2				3	8	1	4	2	7	9	5	6
3	2	5	9	6	1	7	4	8				6	5	4	9	3	1	7	8	2
1	5	7	8	4	2	6	3	9	5	1	7	4	2	8	7	1	3	5	6	9
4	8	6	1	3	9	2	7	5	3	8	4	9	1	6	5	4	2	8	3	7
2	9	3	7	5	6	8	1	4	6	2	9	5	3	7	8	6	9	2	4	1
						7	8	3	9	4	2	1	6	5						
						9	5	1	8	6	3	7	4	2						
						4	6	2	1	7	5	8	9	3						
9	5	7	6	1	2	3	4	8	2	5	1	6	7	9	8	4	2	5	3	1
4	8	3	7	9	5	1	2	6	7	9	8	3	5	4	1	7	6	9	8	2
6	2	1	8	3	4	5	9	7	4	3	6	2	8	1	9	3	5	6	4	7
7	3	2	1	4	8	9	6	5				8	3	6	7	5	1	2	9	4
1	6	5	3	7	9	2	8	4				9	4	7	6	2	8	3	1	5
8	4	9	2	5	6	7	1	3				1	2	5	3	9	4	8	7	6
2	1	6	5	8	3	4	7	9				4	9	3	2	6	7	1	5	8
5	7	4	9	6	1	8	3	2				7	1	2	5	8	9	4	6	3
3	9	8	4	2	7	6	5	1				5	6	8	4	1	3	7	2	9

91

4	3	9	5	8	6	1	7	2				5	3	9	4	6	7	1	2	8
5	2	8	7	3	1	4	9	6				2	7	1	3	8	5	9	6	4
6	7	1	2	4	9	3	8	5				8	6	4	2	9	1	5	3	7
3	9	4	6	1	7	5	2	8				4	5	2	8	3	9	6	7	1
1	6	2	4	5	8	7	3	9				9	8	7	5	1	6	3	4	2
8	5	7	3	9	2	6	1	4				6	1	3	7	4	2	8	9	5
7	4	3	9	2	5	8	6	1	7	2	9	3	4	5	9	2	8	7	1	6
2	1	5	8	6	3	9	4	7	5	6	3	1	2	8	6	7	3	4	5	9
9	8	6	1	7	4	2	5	3	8	4	1	7	9	6	1	5	4	2	8	3
						7	2	8	1	5	6	4	3	9						
						6	1	4	9	3	2	8	5	7						
						5	3	9	4	7	8	6	1	2						
2	9	7	1	3	5	4	8	6	3	9	5	2	7	1	8	9	4	6	5	3
1	6	5	8	7	4	3	9	2	6	1	7	5	8	4	6	1	3	7	9	2
8	4	3	2	9	6	1	7	5	2	8	4	9	6	3	5	2	7	1	4	8
3	2	4	5	1	7	9	6	8				8	2	6	3	5	9	4	7	1
7	8	1	4	6	9	2	5	3				7	3	5	4	6	1	2	8	9
6	5	9	3	8	2	7	4	1				1	4	9	7	8	2	3	6	5
4	3	6	9	2	8	5	1	7				6	1	2	9	4	8	5	3	7
9	7	2	6	5	1	8	3	4				4	9	7	1	3	5	8	2	6
5	1	8	7	4	3	6	2	9				3	5	8	2	7	6	9	1	4

92

9	2	4	1	7	3	6	5	8				1	9	4	6	2	8	7	5	3
5	7	8	4	9	6	1	2	3				6	8	3	9	7	5	2	4	1
1	6	3	5	2	8	7	9	4				5	2	7	3	4	1	6	9	8
4	5	7	6	8	2	3	1	9				7	3	2	1	9	6	4	8	5
3	1	9	7	4	5	8	6	2				8	6	9	7	5	4	1	3	2
2	8	6	9	3	1	5	4	7				4	1	5	2	8	3	9	7	6
6	9	2	8	1	7	4	3	5	8	2	6	9	7	1	8	3	2	5	6	4
8	3	1	2	5	4	9	7	6	5	1	3	2	4	8	5	6	7	3	1	9
7	4	5	3	6	9	2	8	1	9	7	4	3	5	6	4	1	9	8	2	7
						5	9	3	1	8	7	6	2	4						
						7	4	8	3	6	2	1	9	5						
						6	1	2	4	5	9	7	8	3						
4	3	1	2	6	7	8	5	9	2	3	1	4	6	7	1	8	9	3	5	2
6	5	9	8	3	4	1	2	7	6	4	8	5	3	9	2	4	7	6	8	1
7	2	8	1	9	5	3	6	4	7	9	5	8	1	2	6	3	5	7	4	9
9	4	7	6	1	8	5	3	2				9	8	1	3	5	2	4	6	7
3	8	5	4	7	2	9	1	6				7	4	5	8	6	1	9	2	3
2	1	6	3	5	9	4	7	8				6	2	3	7	9	4	5	1	8
1	6	4	9	2	3	7	8	5				2	7	4	5	1	3	8	9	6
5	9	3	7	8	6	2	4	1				1	9	6	4	7	8	2	3	5
8	7	2	5	4	1	6	9	3				3	5	8	9	2	6	1	7	4

SOLUTIONS

93

3	2	5	4	6	9	1	8	7				9	4	6	7	8	5	2	1	3
6	8	1	7	2	5	9	4	3				8	5	1	2	6	3	7	9	4
7	9	4	3	1	8	6	2	5				2	7	3	9	1	4	8	5	6
4	3	2	8	5	1	7	6	9				6	1	8	4	3	9	5	2	7
5	7	8	9	4	6	2	3	1				7	3	4	8	5	2	9	6	1
9	1	6	2	7	3	8	5	4				5	2	9	6	7	1	3	4	8
2	6	7	1	3	4	5	9	8	4	2	3	1	6	7	5	2	8	4	3	9
8	5	3	6	9	7	4	1	2	8	6	7	3	9	5	1	4	7	6	8	2
1	4	9	5	8	2	3	7	6	9	5	1	4	8	2	3	9	6	1	7	5
						8	5	1	7	4	6	2	3	9						
						6	3	7	1	9	2	5	4	8						
						2	4	9	3	8	5	7	1	6						
7	9	6	5	8	4	1	2	3	6	7	9	8	5	4	1	7	2	9	6	3
1	2	4	7	6	3	9	8	5	2	1	4	6	7	3	4	8	9	2	5	1
8	5	3	1	9	2	7	6	4	5	3	8	9	2	1	3	5	6	4	8	7
2	3	1	4	5	9	6	7	8				5	3	8	9	6	1	7	2	4
5	4	8	6	2	7	3	1	9				4	6	7	8	2	3	1	9	5
9	6	7	8	3	1	4	5	2				1	9	2	7	4	5	8	3	6
3	8	9	2	1	6	5	4	7				2	4	6	5	1	8	3	7	9
6	7	2	9	4	5	8	3	1				7	8	9	6	3	4	5	1	2
4	1	5	3	7	8	2	9	6				3	1	5	2	9	7	6	4	8

94

5	3	8	2	4	1	9	7	6				1	3	6	2	9	5	8	7	4
4	9	6	3	7	8	5	1	2				9	4	7	3	8	1	2	5	6
7	1	2	9	6	5	3	4	8				2	5	8	6	4	7	1	9	3
6	2	4	8	3	7	1	5	9				7	2	5	4	1	8	3	6	9
3	7	1	6	5	9	8	2	4				6	9	3	5	7	2	4	1	8
8	5	9	1	2	4	6	3	7				8	1	4	9	6	3	7	2	5
2	6	7	5	8	3	4	9	1	8	6	5	3	7	2	8	5	9	6	4	1
9	4	5	7	1	6	2	8	3	9	7	4	5	6	1	7	3	4	9	8	2
1	8	3	4	9	2	7	6	5	3	1	2	4	8	9	1	2	6	5	3	7
						8	1	9	7	3	6	2	5	4						
						5	7	4	2	9	8	1	3	6						
						3	2	6	4	5	1	7	9	8						
4	7	3	8	1	9	6	5	2	1	8	7	9	4	3	2	1	6	8	5	7
2	6	1	3	5	7	9	4	8	5	2	3	6	1	7	9	8	5	4	3	2
9	8	5	2	4	6	1	3	7	6	4	9	8	2	5	7	3	4	1	9	6
6	3	9	7	2	8	4	1	5				2	3	1	4	6	9	7	8	5
1	5	2	6	3	4	8	7	9				5	9	8	1	7	2	3	6	4
7	4	8	1	9	5	2	6	3				7	6	4	3	5	8	2	1	9
3	9	4	5	6	2	7	8	1				3	5	6	8	2	7	9	4	1
8	1	6	9	7	3	5	2	4				1	7	9	5	4	3	6	2	8
5	2	7	4	8	1	3	9	6				4	8	2	6	9	1	5	7	3

95

2	6	1	3	8	5	4	7	9				7	1	6	4	5	2	8	3	9
3	9	7	6	2	4	1	5	8				9	5	2	6	8	3	7	4	1
5	8	4	9	7	1	2	3	6				4	8	3	7	9	1	5	6	2
6	3	2	5	9	8	7	1	4				5	6	1	9	3	7	4	2	8
1	7	9	4	6	3	8	2	5				3	2	4	8	6	5	1	9	7
4	5	8	2	1	7	9	6	3				8	9	7	1	2	4	6	5	3
9	4	5	1	3	2	6	8	7	9	3	2	1	4	5	3	7	9	2	8	6
7	1	6	8	5	9	3	4	2	5	8	1	6	7	9	2	4	8	3	1	5
8	2	3	7	4	6	5	9	1	7	4	6	2	3	8	5	1	6	9	7	4
						2	7	4	3	6	5	9	8	1						
						9	3	5	2	1	8	7	6	4						
						1	6	8	4	9	7	5	2	3						
2	9	8	1	3	7	4	5	6	1	2	3	8	9	7	1	3	2	4	5	6
4	7	6	9	5	2	8	1	3	6	7	9	4	5	2	9	6	8	1	3	7
3	5	1	8	4	6	7	2	9	8	5	4	3	1	6	4	7	5	2	9	8
6	1	2	3	9	4	5	8	7				2	3	1	6	8	4	9	7	5
8	4	7	5	6	1	3	9	2				7	8	4	3	5	9	6	2	1
9	3	5	7	2	8	6	4	1				5	6	9	7	2	1	8	4	3
5	8	9	6	1	3	2	7	4				9	7	3	2	1	6	5	8	4
7	6	4	2	8	9	1	3	5				6	4	8	5	9	3	7	1	2
1	2	3	4	7	5	9	6	8				1	2	5	8	4	7	3	6	9

96

6	8	1	5	7	3	9	4	2				5	4	9	3	6	2	7	1	8
2	7	5	1	4	9	8	3	6				3	8	7	5	1	9	6	2	4
9	4	3	2	6	8	5	1	7				6	1	2	8	4	7	3	5	9
8	6	9	4	3	2	1	7	5				1	9	5	4	7	6	2	8	3
1	2	7	6	8	5	3	9	4				2	7	6	9	3	8	5	4	1
5	3	4	7	9	1	2	6	8				8	3	4	1	2	5	9	7	6
3	1	6	8	2	4	7	5	9	2	3	1	4	6	8	2	5	3	1	9	7
7	5	2	9	1	6	4	8	3	9	5	6	7	2	1	6	9	4	8	3	5
4	9	8	3	5	7	6	2	1	4	7	8	9	5	3	7	8	1	4	6	2
						3	6	7	5	2	4	8	1	9						
						1	4	2	8	9	3	5	7	6						
						5	9	8	1	6	7	2	3	4						
2	6	4	9	7	3	8	1	5	6	4	2	3	9	7	5	2	1	4	6	8
5	8	7	2	6	1	9	3	4	7	1	5	6	8	2	4	7	9	1	5	3
9	3	1	5	8	4	2	7	6	3	8	9	1	4	5	3	6	8	2	9	7
7	2	3	1	9	6	5	4	8				5	3	4	6	9	2	7	8	1
1	5	9	4	3	8	7	6	2				8	1	9	7	4	5	3	2	6
8	4	6	7	5	2	1	9	3				7	2	6	1	8	3	5	4	9
3	7	8	6	1	5	4	2	9				2	7	3	9	5	6	8	1	4
4	1	5	3	2	9	6	8	7				4	6	8	2	1	7	9	3	5
6	9	2	8	4	7	3	5	1				9	5	1	8	3	4	6	7	2

SOLUTIONS

97

3	2	4	9	6	1	8	5	7				8	3	6	2	5	7	1	4	9
5	7	1	8	4	2	3	6	9				9	5	1	4	6	8	2	3	7
6	9	8	7	5	3	2	4	1				7	2	4	9	1	3	6	8	5
7	1	3	6	2	5	9	8	4				5	4	7	8	2	9	3	1	6
8	4	5	1	9	7	6	2	3				2	6	9	3	7	1	4	5	8
2	6	9	3	8	4	7	1	5				1	8	3	5	4	6	7	9	2
4	3	6	5	7	8	1	9	2	5	6	4	3	7	8	1	9	2	5	6	4
1	8	2	4	3	9	5	7	6	1	8	3	4	9	2	6	3	5	8	7	1
9	5	7	2	1	6	4	3	8	9	7	2	6	1	5	7	8	4	9	2	3
						3	1	4	6	2	8	7	5	9						
						7	8	9	3	1	5	2	6	4						
						6	2	5	7	4	9	1	8	3						
4	5	7	2	1	8	9	6	3	2	5	1	8	4	7	6	9	5	3	2	1
2	9	3	4	7	6	8	5	1	4	3	7	9	2	6	7	3	1	4	5	8
6	1	8	9	3	5	2	4	7	8	9	6	5	3	1	2	8	4	6	7	9
7	4	6	1	8	9	3	2	5				7	6	4	9	2	3	1	8	5
3	8	1	7	5	2	4	9	6				2	5	3	8	1	7	9	6	4
5	2	9	6	4	3	1	7	8				1	9	8	5	4	6	2	3	7
8	3	4	5	9	7	6	1	2				3	7	9	1	6	8	5	4	2
9	6	5	3	2	1	7	8	4				6	8	2	4	5	9	7	1	3
1	7	2	8	6	4	5	3	9				4	1	5	3	7	2	8	9	6

98

3	8	4	1	9	2	5	6	7				1	9	3	4	8	7	2	5	6
5	9	6	7	4	3	2	1	8				6	2	8	5	3	1	9	4	7
1	2	7	8	5	6	4	9	3				5	4	7	6	2	9	8	1	3
7	3	9	2	1	8	6	5	4				4	6	1	2	9	3	5	7	8
8	1	5	6	7	4	3	2	9				3	8	9	1	7	5	4	6	2
6	4	2	5	3	9	8	7	1				7	5	2	8	4	6	3	9	1
4	6	8	9	2	7	1	3	5	8	4	9	2	7	6	9	5	8	1	3	4
2	7	1	3	8	5	9	4	6	2	7	1	8	3	5	7	1	4	6	2	9
9	5	3	4	6	1	7	8	2	5	6	3	9	1	4	3	6	2	7	8	5
						3	7	9	6	5	2	1	4	8						
						5	6	1	3	8	4	7	2	9						
						4	2	8	1	9	7	5	6	3						
3	4	6	9	1	2	8	5	7	4	1	6	3	9	2	7	1	4	5	8	6
9	7	8	6	3	5	2	1	4	9	3	8	6	5	7	3	9	8	2	4	1
1	5	2	4	8	7	6	9	3	7	2	5	4	8	1	6	5	2	3	7	9
5	8	1	3	9	4	7	6	2				5	1	6	8	3	7	4	9	2
7	3	9	2	6	1	4	8	5				2	3	8	4	6	9	1	5	7
2	6	4	7	5	8	1	3	9				7	4	9	5	2	1	8	6	3
6	2	5	1	4	3	9	7	8				1	7	4	2	8	6	9	3	5
8	9	7	5	2	6	3	4	1				9	6	3	1	4	5	7	2	8
4	1	3	8	7	9	5	2	6				8	2	5	9	7	3	6	1	4

99

8	1	3	2	7	6	9	5	4				9	3	2	4	7	5	8	6	1
4	9	6	1	5	3	2	7	8				1	5	4	8	3	6	9	7	2
5	7	2	9	4	8	6	3	1				7	8	6	1	2	9	4	5	3
3	2	7	8	6	1	4	9	5				4	6	7	9	1	2	5	3	8
6	4	1	7	9	5	3	8	2				8	2	3	6	5	4	7	1	9
9	5	8	4	3	2	7	1	6				5	9	1	7	8	3	2	4	6
2	3	5	6	1	7	8	4	9	3	7	2	6	1	5	2	9	7	3	8	4
1	6	9	3	8	4	5	2	7	8	1	6	3	4	9	5	6	8	1	2	7
7	8	4	5	2	9	1	6	3	4	9	5	2	7	8	3	4	1	6	9	5
						9	3	2	6	5	4	7	8	1						
						7	1	6	9	8	3	5	2	4						
						4	8	5	1	2	7	9	3	6						
3	2	1	5	7	4	6	9	8	2	3	1	4	5	7	3	8	6	9	1	2
8	4	5	9	2	6	3	7	1	5	4	9	8	6	2	1	9	5	7	4	3
6	7	9	8	3	1	2	5	4	7	6	8	1	9	3	2	4	7	5	8	6
4	5	2	7	1	8	9	6	3				7	1	6	4	5	2	3	9	8
9	1	8	6	5	3	7	4	2				3	2	8	9	7	1	4	6	5
7	6	3	4	9	2	1	8	5				5	4	9	6	3	8	2	7	1
1	8	7	2	6	5	4	3	9				9	7	1	5	6	3	8	2	4
5	3	6	1	4	9	8	2	7				2	3	4	8	1	9	6	5	7
2	9	4	3	8	7	5	1	6				6	8	5	7	2	4	1	3	9

100

6	2	8	7	5	3	4	1	9				9	4	7	1	6	8	3	2	5
5	4	9	8	6	1	7	2	3				8	3	2	4	9	5	7	1	6
1	3	7	9	2	4	8	5	6				1	6	5	7	2	3	4	8	9
7	6	2	5	8	9	3	4	1				5	2	1	9	8	7	6	3	4
4	9	1	6	3	2	5	7	8				6	8	3	2	5	4	1	9	7
8	5	3	4	1	7	9	6	2				7	9	4	6	3	1	2	5	8
3	7	4	1	9	6	2	8	5	4	1	9	3	7	6	5	1	9	8	4	2
2	8	6	3	4	5	1	9	7	6	3	2	4	5	8	3	7	2	9	6	1
9	1	5	2	7	8	6	3	4	7	5	8	2	1	9	8	4	6	5	7	3
						5	1	9	3	2	7	8	6	4						
						4	7	2	8	9	6	1	3	5						
						8	6	3	5	4	1	7	9	2						
2	1	5	7	8	9	3	4	6	1	8	5	9	2	7	5	4	3	8	1	6
6	9	8	4	3	5	7	2	1	9	6	4	5	8	3	9	6	1	7	4	2
7	3	4	6	2	1	9	5	8	2	7	3	6	4	1	8	7	2	5	9	3
3	6	2	1	5	7	4	8	9				4	6	9	1	8	7	2	3	5
4	8	1	2	9	6	5	3	7				7	5	2	6	3	9	4	8	1
9	5	7	3	4	8	6	1	2				1	3	8	4	2	5	9	6	7
5	2	6	8	7	3	1	9	4				8	1	6	7	5	4	3	2	9
8	7	3	9	1	4	2	6	5				2	9	5	3	1	8	6	7	4
1	4	9	5	6	2	8	7	3				3	7	4	2	9	6	1	5	8

I hope you have enjoyed this book. Please remember to rate it on Amazon! Check out my other graded Killer Sudoku books, for all levels of expertise.

*Killer Sudoku for Beginners, Books 1** and 2*

80 1-star, 80 2-star and 40 3-star puzzles.

Killer Sudoku, Intermediate Level, Book 1

60 2-star puzzles, 80 3-star puzzles and 60 4-star puzzles

*Killer Sudoku for Experts, Books 1, 2**, 3 and 4***

40 3-star puzzles, 80 4-star puzzles and 80 5-star puzzles

*Killer Sudoku, Books 1, 2** and 3*

200 puzzles ranging from 1-star to 5-star

** Also available in large print in A4 format. Puzzle squares are a massive 12.3 cm x 12.3 cm. providing more working space.

If you like regular Sudoku you should challenge the Samurai – each Samurai is five linked Sudoku puzzles to battle against.

Samurai Sudoku, Book 1

100 Samurai 5-in-1 puzzles, with target solving times from 40 minutes to over 2 hours.

All of these books make excellent presents for the Sudoku puzzlers among your friends and family who need a new challenge.

www.ingramcontent.com/pod-product-compliance
Lightning Source LLC
Chambersburg PA
CBHW080835220526
45467CB00008B/2279
9798715201645

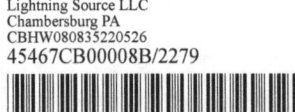